Applied Chemistry of the Alkali Metals

Applied Chemistry of the Alkali Metals

Hans U. Borgstedt
Kernforschungszentrum Karlsruhe
Karlsruhe, Federal Republic of Germany

and

Cherian K. Mathews
Indira Gandhi Center for Atomic Research
Kalpakkam, India

Plenum Press • New York and London

Library of Congress Cataloging in Publication Data

Borgstedt, H. U. (Hans Ulrich), 1930–
Applied chemistry of the alkali metals.

Includes bibliographical references and index.
1. Alkali metals. I. Mathews, Cherian K. II. Title.
QD172.A4B64 1987 546′.38 87-2466
ISBN 0-306-42326-X

A Division of Plenum Publishing Corporation
233 Spring Street, New York, N.Y. 10013

Printed in the United States of America

Preface

Alkali metals constitute a group of elements which are scientifically interesting and technologically important. Their pronounced metallic nature, long liquid range, strong electropositive character, and high chemical reactivity make them unique in their properties. While the study of the chemistry of these elements is interesting in itself, it is also a prerequisite to their commercial and technological applications.

The advent of fast breeder reactors, which employ liquid sodium as a coolant, generated renewed interest in this family of elements. More recently, lithium is finding application in fusion reactors. Both lithium and sodium are used as electrodes in alkali metal batteries. Potassium is used in heat transport systems. Rubidium and cesium are encountered as fission products.

These technological applications have spurred extensive research on alkali metals, resulting in a large body of literature on various aspects of their chemical and physical properties, purification and handling, analytical characterization, and compatibility with materials. However, this literature is largely to be found in the form of conference proceedings, reports, and papers, and these are not convenient forms for the users of alkali metals and researchers in this field. It is in order to fill this gap that we have attempted to put together in one book all the chemical information needed to study and use the alkali metals, particularly in the liquid form.

We have had the opportunity of taking part in the exciting developments in alkali metal chemistry over the past couple of decades. Though set in different environments, one in an industrialized European country and the other in a developing Asian country, we have been involved in the full range of activities necessary for the study of liquid alkali metals. Thus we have set up facilities for handling the liquid metals in the pure form, developed analytical techniques and on-line meters for impurity measurements, measured the solubilities of metals and nonmetals in the liquid metals, and studied chemical reactions in the nonaqueous medium. Our cooperation has also demonstrated that long distances need not stand in the way of collaborative research if there is interest from both sides. In fact, this book is an example of this successful collaboration.

This cooperative effort was nurtured under the aegis of the Indo-German Bilateral Cooperation Agreement, and for this we are thankful to the authorities in both countries. We are also grateful to our employers, Kernforschungszentrum Karlsruhe GmbH, Federal Republic of Germany, and the Indira Gandhi Center for Atomic Research, Kalpakkam, India, for giving permission to undertake this work. It is a pleasure to acknowledge the contributions of our colleagues in the preparation of the manuscript. Messrs. T. Gnanasekaran, S. Rajan Babu, and S. Vanavaraban (Kalpakkam) assisted in finalizing some of the chapters. Miss A. Borgstedt (Karlsruhe) assisted in editing some chapters. Thc authors are also grateful to several publishers for granting permission to make use of material to which they hold copyright (as indicated in the relevant figure captions).

H. U. Borgstedt, *Karlsruhe*
C. K. Mathews, *Kalpakkam*

Contents

Applied Chemistry of the Alkali Metals

1

Introduction

Alkali metals in various forms have been important to man from the dawn of civilization. The ubiquitous common salt has always been a necessary ingredient in the human diet. Soap as well as soda ash, and, even prior to that, ash itself, have been indispensable for washing and cleaning. Natron, an impure form of sodium carbonate, was known in Egypt in very early times. A washing agent called "neter" is referred to in the Old Testament. The word alkali itself is derived from the Arabic *al-qualiy*, calcined ashes. It is, however, the hydroxides of the principal members of the group that have become known under the name "alkalis." Today, caustic soda is one of the major products of the chemical industry.

The six alkali group elements—lithium, sodium, potassium, rubidium, cesium, and francium—constitute the first group of the periodic table. The first five members of the group are naturally occurring, stable elements. The last member of the group, francium, is radioactive and all its isotopes are short-lived. Therefore, it is only of academic curiosity. Sodium is the most common and readily available of all alkali metals. It is the sixth most abundant element and constitutes 2.83% of the earth's crust. In fact, next to aluminum, iron, and calcium, it is the most abundant metallic element. Potassium comes a close second in abundance but is not as widely distributed as sodium. The other alkali metals are much less abundant. The relative scarcity of lithium in the earth's crust is attributed to its disappearance through the reaction

$$ {}^{7}_{3}\mathrm{Li} + {}^{1}\mathrm{H} \rightarrow 2\,{}^{4}_{2}\mathrm{He} \qquad (1) $$

Table 1.1 lists the natural abundances of the alkali metals and the principal minerals from which they are obtained.[1]

Table 1.1. Occurrence of Alkali Metals

Alkali metal	Abundance in earth's crust (in ppm)	Order of abundance	Important minerals bearing alkali metals		Remarks
Lithium	65	27	Spodumene	$LiAlSi_2O_6$ (~8% Li_2O)	Principal commercial source
			Lepidolite (Mica)	$K(Al_2[AlSi_3O_{10}],$ $Li_2[AlSi_3O_6(OH, F)_4])$ $(OH, F)_2$ (4.2–4.5% Li_2O)	May also contain Rb and Cs
			Petallite	$Li_2O \cdot Al_2O_3 \cdot SiO_2$ (3.5–4.5% Li_2O)	
			Amblygonite	$LiAl(F, OH)PO_4$ (~10% Li_2O)	Only a minor quantity occurs
Sodium	28300	6	Rocksalt	$NaCl$	Sea water is the major source
			Borax	$Na_2B_4O_7 \cdot 10H_2O$	
			Soda	$Na_2CO_3 \cdot 10H_2O$	
			Trona	$Na_2CO_3 \cdot NaHCO_3 \cdot 2H_2O$	
			Chile Saltpeter	$NaNO_3$	
			Thenardite	Na_2SO_4	
			Cryolite	Na_3AlF_3	
Potassium	25900	7	Leucite	$K[AlSi_2O_6]$	
			Sylvite	KCl	
			Carnallite	$KCl \cdot MgCl_2 \cdot 6H_2O$	
			Langbeinite	$K_2SO_4 \cdot 2MgSO_4$	
			Polyhalite	$K_2SO_4 \cdot MgSO_4 \cdot 2CaSO_4 \cdot 2H_2O$	
Rubidium	310	16	Lepidolite Carnallite	(~0.24% Rb_2O)	Rb and Cs occur in minute amounts with other alkali metals
Cesium	7	40	Pollucite	$2Cs_2O \cdot 2Al_2O_3 \cdot 9SiO_2 \cdot H_2O$ (10–35% Cs_2O)	
			Rhodizite	$NaKLi_4Be_3B_{10}O_{27}$ (0–3% Cs_2O)	Cs and Rb substitute for K

Note: Abundance data from *Principles of Geochemistry*, Brian Mason, John Wiley and Sons, New York (1952).

As alkali metals are very reactive, they do not occur in nature in the elemental form. The first alkali metal to be isolated was potassium. In 1807, Sir Humphrey Davy passed an electric current through caustic potash using a powerful battery. He found metallic globules of a reactive element collecting at the cathode and called it potassium. A few months later, Davy discovered sodium when he electrolyzed caustic soda. Up to that time, these alkalis had been thought to be elements. Lithium was discovered by Arfvedson in 1818. Working under Berzelius in Stockholm, he found a new constituent element in petalite and called it lithium (from the Greek *lithos*, stone). However, it was left to Brandes and Davy (1820) to isolate small amounts of the metal by the electrolysis of lithia. Larger quantities of metallic lithium were obtained by Bunsen and Matthiessen (1855) by the electrolysis of the fused chloride. Cesium was discovered by Bunsen and Kirchhoff in 1860. Working at Heidelberg with their new spectroscope, they observed the characteristic blue lines of the new element during their studies on mineral waters of Durkheim and named it caesium after the Latin *caesius* (blue sky). They succeeded in isolating 50 g of cesium chloroplatinate from 40 tonnes of mineral waters. Cesium metal was first isolated by Setterberg in 1882 by the electrolysis of its cyanide in the presence of barium cyanide. In 1861, Bunsen and Kirchhoff discovered a new alkali metal in lepidolite. Rubidium derives its name from its characteristic dark red lines (the Latin *rubidus* means dark red). Although several claims were made to the discovery in nature of element 87, the last member of the series, none was substantiated. In 1939, Perey discovered the element as a branch product of natural radioactivity in the actinium series and suggested the name francium.(2)

The applications of the alkali group elements stem from their metallic nature, low melting points, high chemical reactivity, and low ionization potentials. Sodium, the most abundant and cheapest of all alkali metals, is the most important in industrial applications. The first industrial use of it was in the manufacture of pure aluminum by the reduction of aluminum chloride. The Woehler process for producing aluminum, which was developed in 1827, actually used potassium as the reducing agent. This was replaced by sodium in the process developed by Deville and Bunsen in 1854. However, the demand for sodium on this account was short lived because the aluminum industry gradually switched over to the electrolytic process. Sodium soon found application in the manufacture of sodium cyanide and sodium peroxide. The second half of the nineteenth century saw also the widespread use of sodium in organic reactions. The Williamson reaction (1850), Fittig synthesis (1863), Claisen condensation (1863), and Wurtz synthesis (1885) are classical examples. Sodamide is a raw material in the preparation of cyclic compounds, for example, indigo. Thus, in 1927, when the world production of sodium was 27,000 tons, 40% of it was used for peroxide and cyanide, 25% for organic reductions, and an important fraction for sodium amide for use in indigo manufacture. The Second World War led to an increased demand for sodium, especially in the manufacture of tetraethyl lead (TEL) and sodium cyanide. TEL, which is used as an antiknock compound in gasoline, is prepared by the action of ethyl chloride on sodium–lead alloy. Towards the end of the fifties, when the world production of sodium stood at around 200,000 tonnes, about half that quantity was used in TEL manufacture.[3]

Alkali metals also find use in illuminating engineering, the most common example being the sodium vapor discharge lamp. Potassium and especially cesium are used in photoelectric cells. The photocathode, the basic element of a photomultiplier, usually contains cesium. Addition of lithium improves the properties of many technical alloys to a notable degree. Lithium is used in lead base bearing metals and in aluminum alloys.

1.1. Alkali Metals as Heat Transfer Media

Applications of alkali metals are based on their physical and chemical properties and their availability. Their properties are discussed in Chapter 2 and their production in Chapter 3. Because of their excellent heat transfer properties, alkali metals find many applications in engineering. Utilization of these metals in conventional applications is described by Katz[4] and by Jackson.[5] The use of liquid sodium for cooling valves of combustion engines and plows in shale-oil retorts has been reported. Other conventional uses include the use of sodium-filled cores for die casting and NaK, the eutectic sodium–potassium alloy, for heat treatment furnaces.

Today, there is renewed interest in alkali metals because of their tremendous potential in the production and storage of energy. The energy crisis of the seventies triggered by a sharp rise in the cost of crude oil has accelerated the search for new sources of energy to take the place of fossil fuels, which are fast depleting. If nuclear fission is to provide an alternate source of energy which will last for some centuries, then the uranium resources must be better utilized than is possible through the use of the current generation of thermal reactors. Only fast breeder reactors can provide the answer. There is also considerable effort underway to tap solar energy, which is abundantly available. When nuclear fusion is harnessed for producing electrical energy—perhaps in the next century—it will provide us with another long-lasting source of energy. It is indeed likely that a combination of all these sources may be used in the future. Part of the answer to the energy problems lies in improving energy conversion efficiency by extracting energy at higher temperatures. For effective utilization of intermittent sources of energy and to take care of nonuniform demand, which peaks during some parts of the day, energy storage devices will be required. In all these applications, alkali metals are likely to play a major role. Alkali metals are destined to be the heat transfer medium of the future. It is, therefore, considered appropriate to discuss briefly at this point these newly emerging applications of liquid alkali metals.

Water has traditionally been used as a heat transfer medium in power plants. Even in atomic power plants, water is the coolant, though in some natural uranium reactors heavy water has to be used. For the cooling of the recently developed fast breeder reactors, which have high specific power and, therefore, require a very efficient coolant, liquid sodium was chosen. The alkali metal melt is not only an efficient cooling medium, but it also does not act as a "moderator," as water does, slowing down the fast neutrons and thus making them unavailable for the breeding process.

A good heat transfer medium for nuclear applications must have the following characteristics:

1. High thermal conductivity
2. High specific heat
3. Low viscosity
4. Low melting point and high boiling point so that the coolant is a low-pressure fluid in the operating temperature range
5. Stability to heat and to radiation
6. Compatibility with structural materials
7. Low cost
8. Nonhazardous nature (toxicity, induced radioactivity)
9. Favorable nuclear properties

It is, of course, impossible to find a coolant that meets all these requirements fully. In actual practice, the choice represents a compromise among conflicting requirements. Liquid sodium comes nearest to meeting most of the above conditions, as can be seen from its properties (see Chapter 2). Both lithium and sodium have good thermal properties. Lithium has the highest specific heat of all elements besides hydrogen and helium. Sodium has the higher thermal conductivity and density so that the heat transfer properties of these elements are comparable. Sodium is more readily available and is cheaper than lithium by a factor of 30.

1.2. Fast Breeder Reactors

In a fast breeder reactor, energy is produced by nuclear fission induced by "fast" neutrons. They are called breeder reactors because they are able to produce more fissile materials than they "burn" to produce energy. In thermal reactors, energy is produced through nuclear fission caused by thermal or slow neutrons. It is the fissile isotope of uranium (^{235}U) that fissions. As natural uranium contains only 0.71% ^{235}U (the rest being the nonfissile isotope ^{238}U), it is clear that thermal reactors can extract only a small percentage of the energy available in uranium. The FBRs use a mixture of uranium and plutonium as fuel and it is the plutonium that mainly undergoes fission. The rate at which ^{238}U is converted to ^{239}Pu is higher than the rate at which ^{239}Pu fissions. Thus, FBRs breed more plutonium than they use up for generating energy. Therefore, through FBRs, we can use practically all the uranium (and not just ^{235}U), thus extending our energy supply by almost two orders of magnitude. When fast breeders are further developed to use thorium (by making use of the $^{232}Th/^{233}U$ cycle) we will have an enormous source for the supply of energy by the fission process alone.

Sodium is the coolant of choice for fast reactors. Lithium cannot be used because of unfavorable nuclear properties, high cost, and corrosion

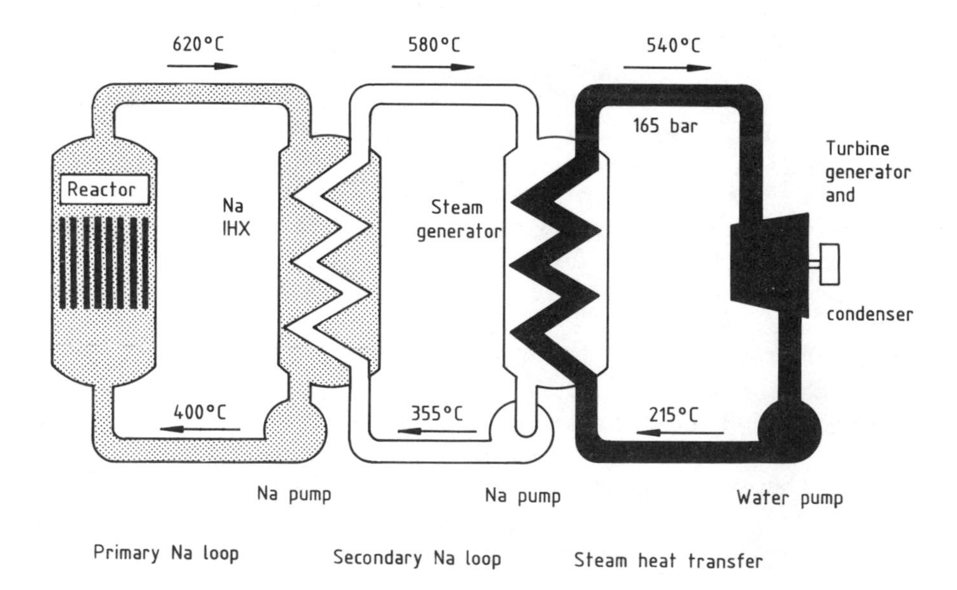

Figure 1.1. Schematic of a liquid-metal-cooled fast breeder reactor.

potential. Figure 1.1 gives a schematic representation of the sodium circuits in a liquid-metal-cooled fast breeder reactor (LMFBR). The heat generated in the reactor core is transported by the primary sodium to the intermediate heat exchanger (IHX). Only in the primary system can the coolant become activated, and this part of the system is confined within the heavily shielded reactor vault. The secondary sodium carries the heat from the IHX to the steam generator and the superheater. This sodium does not get activated. The world's largest fast reactor, the Superphenix in France, is designed to generate 1242 MW of electricity. It contains 3250 tonnes of sodium in the primary system and 1500 tonnes in the secondary systems. The maximum temperature of the sodium is 545°C.[6]

1.3. Fusion Reactors

Nuclear fusion promises to make available another large source of energy. The nuclear fusion reaction between deuterium (D) and tritium (T), the two heavier isotopes of hydrogen, is the basis for the energy from fusion

$$\mathrm{D + T \rightarrow He + n + 17.6\ MeV} \tag{2}$$

Deuterium and tritium nuclei react (or "fuse" together) at very high temperatures to form a helium nucleus and a neutron, generating in this process a lot of energy. Deuterium is obtained from nature in the same way as heavy water is separated from water for use in natural uranium reactors. But because there is no significant supply of tritium, a D–T fusion reactor must breed its own tritium. This is where lithium plays its role in the fusion technology. Natural lithium has two isotopes—^{6}Li and ^{7}Li, with an abundance of 7.4% and 92.6%, respectively. Both these isotopes can produce tritium when bombarded with neutrons.

$$\mathrm{^{6}Li + n \rightarrow {}^{4}He + T} \tag{3}$$

$$\mathrm{^{7}Li + n \rightarrow {}^{4}He + T + n} \tag{4}$$

The former reaction has a high cross section (probability of reaction) at thermal energies whereas the latter is possible only at neutron energies greater than 4 MeV and has a much lower cross section. Nevertheless, the ^{7}Li reaction is also important.

(a)

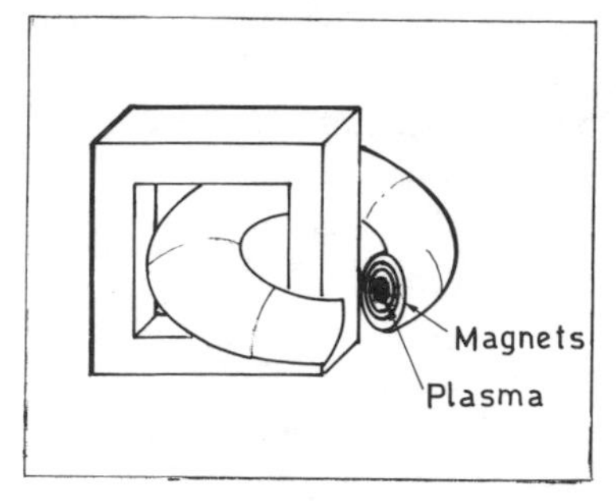

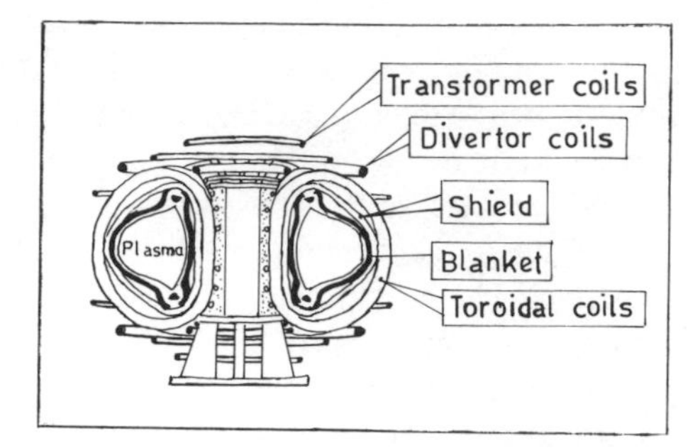

(b)

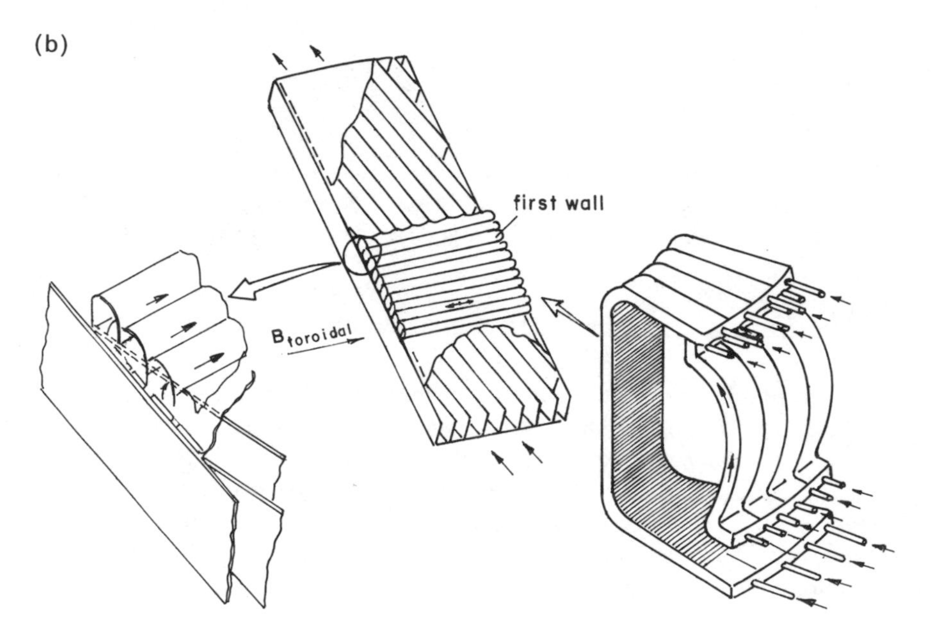

Figure 1.2. (a) Schematic of a tokamak reactor. (b) The ANL fusion blanket concept (after ref. 7).

Inertial confinement and magnetic confinement are the two concepts being pursued in fusion research. In the former, pellets of D–T are bombarded from all sides by laser beams acting in unison. The resultant implosion generates the conditions for fusion. In the latter method, the fusion plasma is confined magnetically. Currently, the magnetic confinement system that is considered most promising is the closed toroidal system known as the tokamak. In both cases, lithium may be thought of as surrounding the reaction zone as a blanket in which breeding of tritium takes place.

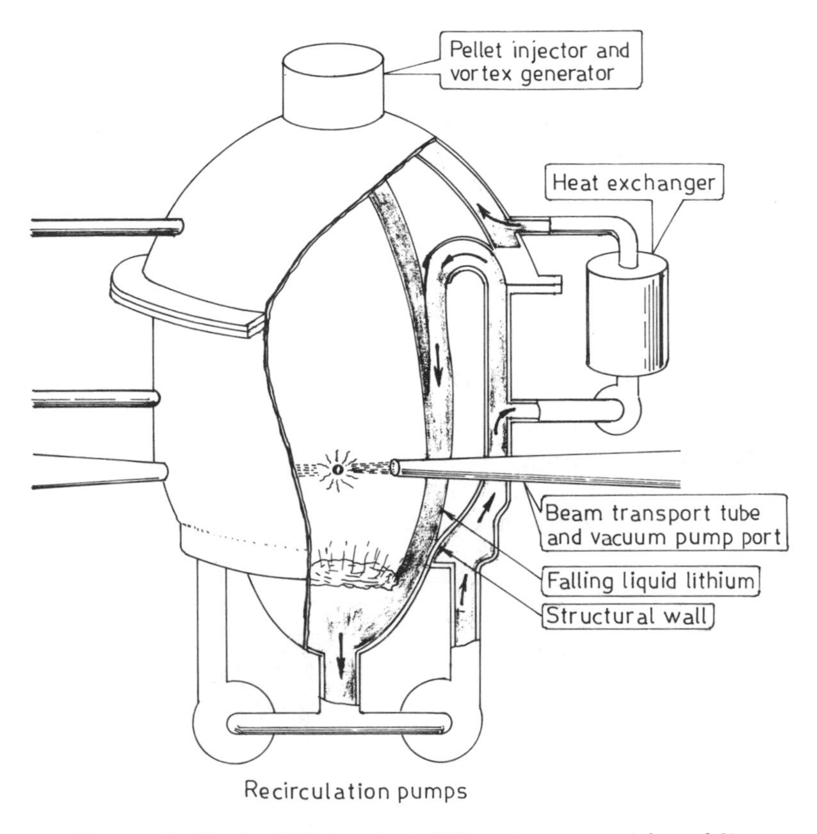

Figure 1.3. The liquid lithium "waterfall" reactor concept (after ref. 8).

Lithium and the eutectic lithium–lead alloy $Li_{17}Pb_{83}$ are attractive blanket materials. The latter has the potential of giving a higher breeding ratio through neutron multiplication. In some tokamak fusion reactor concepts, these liquid metals are used only as blankets while helium or water is used as coolant and transports not only heat but also the tritium produced. However, there are concepts where the liquid metal serves both as breeder material and as coolant. The advantages of such a self-cooled concept are simple mechanical design and ease of tritium recovery. The main disadvantages are: (1) the high pressure drop resulting from liquid metal flow in the high magnetic field (MHD loss) and (2) corrosion of structural materials by the liquid metal. The MHD pressure drop problem is overcome in a recent concept from Argonne National Laboratory (ANL).[7] Here the liquid metal coolant flows in the poloidal direction (perpendicular to the magnetic field) with a low velocity, thus minimizing MHD losses. The higher velocity necessary to cool the first wall is achieved by changing the flow from the poloidal to the toroidal direction through short and narrow tubes. Figure 1.2a shows a section of a torus and Fig. 1.2b illustrates the ANL concept. In one of the concepts for the inertial confinement fusion reactor, falling lithium is also being considered as a first-wall material, which will be subjected to high temperature and radiation (Fig. 1.3).[8] Normal structural materials will not be able to withstand such severe environments for long periods.

1.4. Solar Energy

In the central station concept for electricity production using solar energy, as illustrated in Fig. 1.4, liquid metals can be advantageously used in energy absorption, energy transport, and energy storage. The central receiver of a solar power plant collects the energy reflected to it by an array

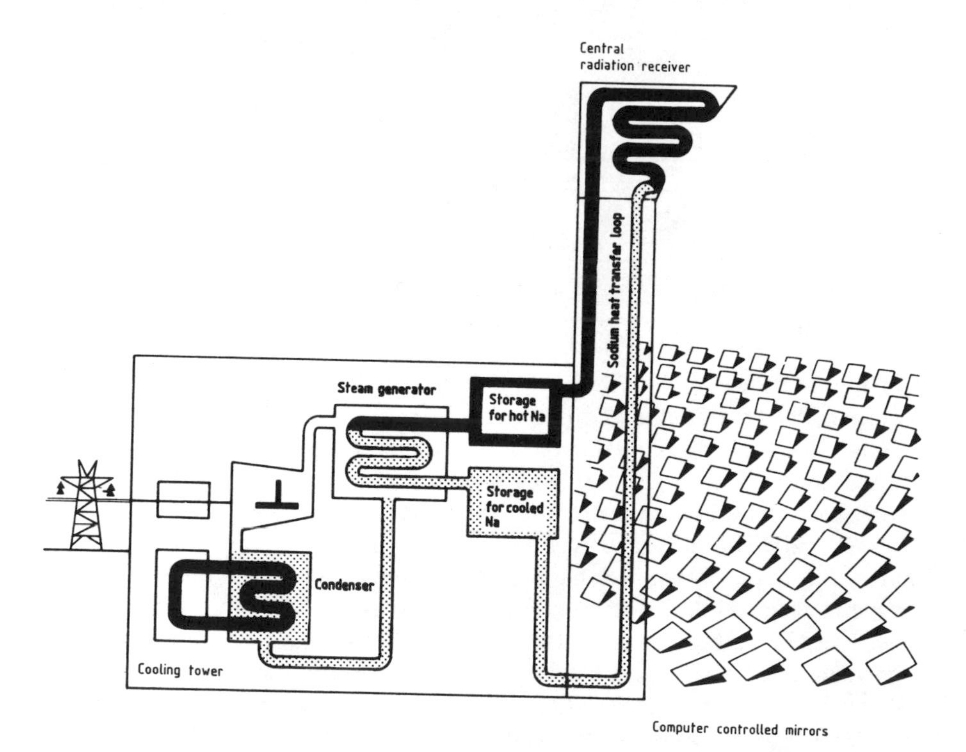

Figure 1.4. The concept of a solar energy power plant using sodium for energy storage and transfer.

of mirrors located around it.[9] This energy has to be absorbed in a suitable medium, which also serves as coolant for the receiver. A liquid metal like sodium is an ideal choice as a receiver coolant because of its excellent heat transfer properties as well as the fact that it remains a liquid throughout the entire temperature range of interest. A solar power station must provide for some temporary energy storage. Even if extension of operation beyond daylight hours is not envisaged, sufficient storage capacity must be available to mitigate the effects of passing clouds. Storage media being considered are rocks, salts, and molten sodium. There are several advantages in using sodium as the storage medium. When sodium is the heat transfer as well as the storage medium, the system becomes compact and simple, and capable of providing steam to a turbine at temperatures and pressures commensurate with modern steam plant requirements.

1.5. Energy Conversion

Improving the thermal efficiency of new central power stations is one approach to solving the energy problem. Studies are underway to develop a new type of central power station which converts thermal energy to electricity at about 60% efficiency, as against the 30 to 40% efficiency available from current-day power stations. To achieve this objective, one must have power plants (e.g., nuclear reactors) which will supply thermal energy at about 950°C to the energy conversion process. Electricity will be produced in two or more Rankine cycles operating with different fluids. In such a scheme, the topping cycle is proposed to be operated using potassium as the working fluid. Potassium has a boiling point of 760°C, which falls within the temperature range of the topping cycle so that its latent heat of evaporation (496 cal/g) adds to the conversion efficiency.[10]

The use of liquid metal heat transfer media to heat up air for the operation of air turbines in the temperature range of 800–1000°C in power plants which combine such high-temperature air turbines with steam engines operating downstream has been proposed. The temperature difference between the heat source, which is typically a coal fire, and the steam at the outlet of the steam generator or superheater is too large so that a sodium cycle leads to a high efficiency of the power plant of the order of 5%. Similar sodium circuits may be used to transfer heat from a high-temperature (helium-cooled) reactor for several process applications. Even in this high-temperature range, sodium circuits are operated at low pressures—below 10 bars.[11]

Another approach to improving the conversion efficiency is to use a magnetohydrodynamic (MHD) power generator cycle. Here, an electrically conducting fluid passes through a magnetic field, thus converting part of the fluid energy directly into electricity. The fluid passing through the MHD duct may be a gas seeded with low-ionizing metal atoms (such as Na or K) or a two-phase fluid consisting of a liquid metal and a gas. The liquid metal–gas combinations considered include NaK–N_2, K–He, K–Ar, Na–He, Na–Ar, and Li–Ar.[12]

Alkali metal energy conversion devices have been proposed which take advantage of the metal-ion-conducting properties of some solid electrolytes. For instance, if a sodium-ion-conducting electrolyte separates two chambers of sodium at different thermodynamic activity levels, then an EMF is generated between these chambers. The open-circuit potential difference between the sodium chambers is a measure of the activity or vapor pressure difference between them. Such a difference in vapor pressures can be maintained if the temperature of sodium in the two chambers is different. If sodium is pumped from the low-pressure side to the high-pressure side by an electromagnetic pump, then the device will operate as a thermoelectric heat engine. It has been suggested that these devices could increase the efficiency of electricity generation by extracting energy at temperatures above the input temperatures of the steam turbine in an analogous way to the proposed operation of MHD generators.[13]

1.6. Energy Storage Devices

Alkali metals have a role to play not only in the production of energy but also in its storage. Lithium-based primary batteries are now widely used, particularly in the electronics industry. Alkali metal-based secondary batteries show great promise of high performance for both electric-driven vehicle and stationary storage applications. Vehicle traction requires a battery with high specific energy. High-capacity batteries are required as load-leveling batteries, which can be available as “spinning reserves” during peak period and which get charged during the slack period. Alkali metal-based batteries with specific energies of about 300 Wh/kg, as against a value of 15 Wh/kg for the common lead–acid battery, are being developed.

The leading secondary batteries under development are the sodium–sulfur and the lithium–sulfur batteries. The former is based on a sodium-ion-conducting electrolyte, β-alumina, which separates a liquid sodium anode and a sulfur/sodium polysulfide cathode. No solid electrolyte is, however, known which conducts the lithium ion. Therefore, lithium batteries under development are based on either liquid eutectics involving a lithium salt or polymer films loaded with a lithium salt. In either case, a solid alloy of lithium constitutes the anode and a suitable sulfide the cathode.

1.7. Heat Pipes

The high boiling temperatures of alkali metals together with their high latent heats of vaporization are made use of in heat pipes to transfer heat. Heat pipes are closed, evacuated tubes of stainless steel or high-temperature alloys containing some amounts of a fluid, usually an alkali metal. One end of the tube is heated by means of an external heat source. The liquid alkali metal gets evaporated at this end, and the vapor transfers the energy to the other end of the tube where a heat sink causes the condensation of the alkali metal vapor. The heat of condensation is transferred to the heat sink (see Fig. 1.5a). The condensed alkali metal is moved back to the heat source by means of the capillarity of the tube wall structure, which is provided with longitudinal narrow channels. These channels act as capillaries and force the liquid metal in a direction opposite to that of the vapor flow.

The alkali metals potassium, sodium, and lithium offer a large range of temperature in which heat pipes might be used. In Figure 1.5b the ranges in which alkali metals might be used as heat transfer media are compared with those of other liquids. The corrosive nature of lithium may, however, limit the temperature range of its application to about 1000°C.

Heat pipes not only have the capability to transfer large quantities of heat but also can transfer heat across small temperature gradients. Therefore, such heat pipes may find applications in heat treatment chambers of large dimensions.

1.8. Chemical Reactivity

All alkali metals are chemically very reactive. They react with oxygen and water with varying degrees of severity depending on the element and the temperature. Sodium quickly tarnishes in air, whereas rubidium and cesium ignite spontaneously. All the metals react vigorously with air and water in the molten state. Lithium reacts also with nitrogen when contacted with air. Because of this reactivity, special precautions have to be taken in handling alkali metals, especially in the liquid state which is more relevant to heat transfer applications.

The complexity of the handling system depends on the application.

(a)

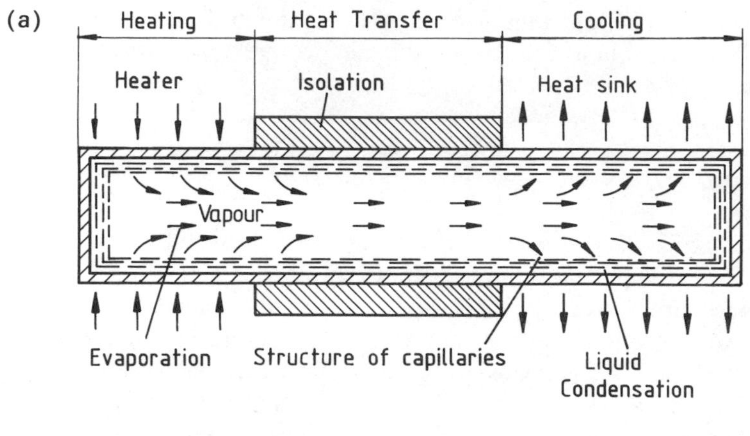

(b)

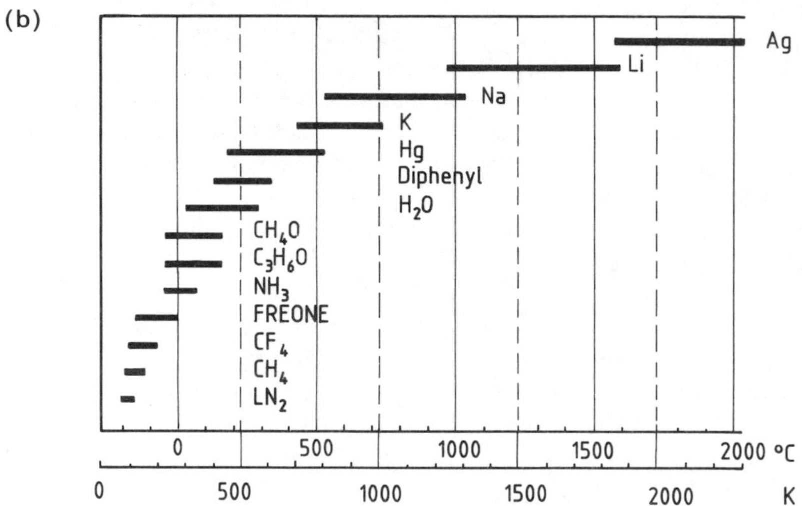

Figure 1.5. (a) Principle of a heat pipe (after ref. 14; reproduced with the permission of Verlag Resch KG, Gräfelfing, Germany). (b) Operating temperature ranges of several heat pipe fluids (after ref. 14; reproduced with the permission of Verlag Resch KG, Gräfelfing, Germany).

For most conventional applications, sodium can be handled in the open air at room temperature as the oxide coating protects it from further reaction with moist air. Precautions are necessary mainly to protect the operator. An inert environment becomes necessary when the temperature is increased. The heavier alkali metals require an inert atmosphere even at room temperature. An argon cover gas is generally used, though nitrogen may be sufficient in some cases. Lithium, however, reacts readily with nitrogen. When alkali metals are used as heat transfer fluids, the level of purity also becomes important. For this reason, the inert atmosphere above the liquid metal must be of very high purity. To achieve this, all valves, joints, and measuring and pumping devices in both the sodium and the cover gas lines must be highly leak-tight. Further, both the sodium and the cover gas lines must be continuously purified by recirculatory purification. Even sampling for chemical analysis requires special devices and extreme care. The techniques necessary for handling, purifying, pumping, and sampling liquid sodium have been developed as part of the liquid-metal-cooled fast breeder reactor technology. They have been adopted for the other alkali metals in part. The various aspects of purification and handling of liquid metals both in the laboratory and in industry are discussed in Chapter 4.

1.9. Corrosion by Liquid Metals and the Role of Impurities

The applications of all liquid alkali metals are limited by the availability of suitable container materials, which withstand corrosion in the temperature and flow velocity ranges of interest. The word "corrosion" is here used in the most general sense—chemical and physical changes due to the effects of the liquid metals on the properties of the container materials.

In the application of liquid sodium as a heat transfer medium, the most commonly used structural material is austenitic stainless steel. Ferritic steels and carbon steels can also be used at lower ranges of temperatures. The heat transfer systems are usually in the form of loops, different parts of which experience different temperatures. The most significant attack on the structural materials is caused by impurities present in sodium, mainly oxygen. Dissolved oxygen affects the system in several ways. The elements of the steels get oxidized at higher temperatures. The oxide layers are not adherent under the influence of flowing alkali metal. This results not only in the depletion of material in the hot part of the circuit but also in the deposition of material in some other (cooler) parts. This type of attack depends on oxygen concentration in sodium, temperature, and temperature gradient. Another effect of oxygen is related to the temperature dependence of its solubility in alkali metals. The solubility of oxygen in sodium varies from 1 ppm at 120°C to 200 ppm at 500°C. Thus, dissolved oxygen precipitates as sodium oxide in the cooler parts of the circuits, causing flow blockage and changes in physical properties of the fluids. It is, therefore, clear that one must understand the behavior of oxygen in liquid alkali metal systems, and this topic is treated in detail in Chapter 5.

Carbon is another impurity whose behavior in liquid alkali metal systems has considerable bearing on the mechanical properties of the container materials. Depending on the chemical activity of carbon in the alkali metal melt, it causes either carburization or decarburization of the structural materials. The resulting carbon transport may be a temperature gradient effect or a concentration gradient effect. The chemistry of carbon in liquid alkali metals is discussed in Chapter 6.

Nitrogen impurity plays a significant role in corrosion by liquid lithium. Weight loss of stainless steel in liquid lithium loops has been correlated with the nitrogen concentration. Grain boundary penetration of steels by lithium is a serious concern when lithium contains higher amounts of nitrogen. These effects can be understood in terms of the solubility and stability of lithium nitride, Li_3N. In the formation of the nitride, lithium resembles more closely the group IIA elements than the other members of the alkali group. Nitrogen also gets transported through sodium to metal surfaces, causing nitriding of solid materials such as stainless steels or refractory metals. The behavior of nitrogen in alkali metals is also discussed in Chapter 6.

Hydrogen is usually present in alkali metals as a result of their reactions with moisture. The solubility of hydrogen in sodium is less than that of oxygen, and hence the hydride precipitates at even higher temperatures. The exchange of hydrogen between alkali metals and solid alloys can cause embrittlement of the materials. Chapter 7 deals with hydrogen in alkali metals. Because of their importance in corrosion and technology, all three elements mentioned above are controlled in modern liquid metal circuits. On-line meters have been developed to monitor their concentration in the alkali metal loops. These meters are also dealt with in the appropriate chapters of the book.

The interactions of oxygen, carbon, hydrogen, and nitrogen with alkali metals have been treated in detail because of their importance in the applications of alkali metals. The solubility of other nonmetals in alkali metals is considered in Chapter 8. These include rare gases which are used as cover gas over liquid alkali metals, halogens, and chalcogens.

Some constituents of structural materials may be leached out by the liquid metals under flowing conditions. At low oxygen levels in sodium loops, chromium and nickel are found to be preferentially leached out from austenitic stainless steel. This is related to the solubility of the metals in liquid sodium. Solubilities of some metals are dependent on oxygen potentials in the liquid metals. Technologically important metals (such as the constituents of the austenitic steels) have very low solubilities in liquid alkali metals. However, solubility may be a strongly varying function of temperature. In such cases, solubility determinations require very careful experimentation. On the other hand, such data are quite important in modeling activity transport, which is caused by corrosion in a fast reactor primary circuit. The alkali metals form alloys and intermetallics with a large number of metals, notably those of groups IB and II. All these interactions are treated in Chapter 9.

After considering the behavior of impurities and the solubilities of elements in liquid alkali metals, we are in a position to discuss in detail the effects of impurities on material compatibility. Thus, in Chapter 10 we present the current understanding of corrosion by liquid alkali metals as well as the effects of alkali metals on the mechanical properties of structural materials.

Liquid alkali metals serve as nonaqueous media for chemical reactions. Formation of nitrides, carbides, and cyanides are examples. Studies of chemical reactions in this medium have the additional attraction that the interpretation of the reaction may be simpler than in the aqueous medium. Chemical reactions in liquid alkali metals are dealt with in Chapter 11. Since sodium–water reactions are important, especially in the steam generator of a fast breeder reactor, this subject is dealt with separately in Chapter 12.

Liquid metal systems make use of components such as pumps, valves, bearing pads etc., in which two surfaces in contact with each other undergo relative motion. Friction and wear of surfaces exposed to liquid alkali metals are markedly different in character compared to those found in normal systems. Conventional lubricants cannot be used in the presence of alkali metals on account of their incompatibility. Unusual phenomena such as self-welding are encountered in liquid metal media. Tribological behaviors of components in liquid metal systems are discussed in Chapter 13.

The various engineering applications of alkali metals such as heat transfer take advantage of their physical properties. The one important application which is based on the chemical properties of the alkali metals is their use in batteries. It is therefore considered appropriate to discuss this aspect in some detail and Chapter 14 is devoted to this purpose.

References

1. H. Remy, *Lehrbuch der Anorganischen Chemie*, Akadem. Verlagsgesellsch., Leipzig, 1965.
2. J. Newton Friend, *Man and the Chemical Elements*, Charles Griffin and Co., London, 1951.
3. M. Sittig, *Sodium—Its Manufacture, Properties and Uses*, Reinhold Publishing Corporation, New York, 1956.

4. D. J. Katz, in: *Liquid Metals Handbook*, (R. N. Lyon, Ed.), NAVEXOS P-733, U.S. Govt. Printing Office, Washington D.C., 1952.
5. C. B. Jackson (Ed.), *Liquid Metals Handbook, Sodium-NaK Supplement*, U.S. Govt. Printing Office, Washington D.C., 1955.
6. Status of Liquid Metal Cooled Fast Breeder Reactors, Technical Report Series No. 243, International Atomic Energy Agency, Vienna, 1985.
7. G. Casani, in: *Liquid Metal Engineering and Technology*, British Nuclear Energy Society, London, 1985, Vol. 3, pp. 303–315.
8. W. R. Meier and J. A. Maniscalco, Reactor Concepts for Laser Fusion, Lawrence Livermore Laboratory Report UCRL-79654, 1977.
9. W. C. Dickinson and P. N. Cheremissinoff (Eds.), *Solar Energy Technology Handbook*, Part B, Marcel Dekker, New York, 1980.
10. G. E. Rajakowics and N. Schwarz, in: *Intern. Conf. on Liquid Metal Technology in Energy Production* (M. H. Cooper, Ed.), National Techn. Information Service, Springfield, Va., 1976, (CONF-760503-P1), pp. 431–436.
11. K. Knizia, *VGB Kraftwerkstechnik 65*, 545–556 (1985).
12. J. A. Shercliff, *A Textbook of Magnetohydrodynamics*, Pergamon Press, Oxford and London, 1965.
13. I. Wynne Jones, in: *Liquid Metal Engineering and Technology*, British Nuclear Energy Society, London, 1985, Vol. 3, pp. 317–321.
14. O. Brost and M. Groll, *Wärme 86*, 36–40, 71–74 (1980).

2

Properties of Alkali Metals

The alkali metal elements, which constitute group IA of the periodic table, are chacterized by their high chemical reactivity, pronounced metallic nature, low melting points, and wide liquid range. This chapter discusses the properties of alkali metals and gives recommended values for important physical properties.

The physical, thermophysical, and chemical properties of these metals determine their applications. Thus, the major technological applications of alkali metals are as heat transfer fluids on account of their excellent heat transfer characteristics and the wide temperature range in which they are liquids. Hence, properties such as melting and boiling points, vapor pressure, density, specific heat, thermal conductivity, viscosity, etc., are given due importance in this chapter. As nuclear properties are relevant in the application of alkali metals in fast breeder reactors and fusion reactors, they are also considered here.

Several compilations of physical properties of alkali metals are available. Some of these are Refs. 1–8 given in the reference list at the end of this chapter. There are, however, very few critical evaluations of alkali metal properties. The *Liquid Metals Handbook* edited by Lyon[2] was published in 1952; since then techniques of alkali metal handling and purification have significantly improved and many new measurements have been carried out. In view of the technological importance of sodium, its properties have been critically reviewed by several authors (see Refs. 4, 7, and 8). Recently an effort was made at the Reactor Research Centre, Kalpakkam, to compare all available literature values and arrive at a set of recommended values.[1] In choosing the recommended values, the authors have kept in mind the experimental conditions such as the purity of the alkali metal and the quality of the atmosphere. Due weight was given to

Table 2.1. Properties of Alkali Metals

Property	Lithium	Sodium	Potassium	Rubidium	Cesium
Atomic number	3	11	19	37	55
Atomic weight	6.941	22.98977	39.0983	85.4678	132.9054
Color and appearance	Silver white, soft	Silver white, soft	Silver white, soft	Silver white, soft	Silver white, soft
Crystal form	BCC	BCC	BCC	BCC	BCC
Lattice constant (Å)	3.50	4.24 (−173°C)	5.31 (−173°C)	5.62 (−173°C)	6.05 (−173°C)
Melting point (K)	453.69	370.98	336.35	312.47	301.67
Melting point (°C)	180.54	97.83	63.20	39.32	28.52
Latent heat of fusion (kJ/g-atom)	3.00	2.597	2.323	2.181	2.092
Volume increase on melting (% of solid volume)	1.5	2.7	2.41	2.54	2.66
Boiling point (K)	1615.15	1154.55	1029	959.15	951.55
Boiling point (°C)	1342	881.4	755.85	686	678.4
Latent heat of vaporization (kJ/g-atom)	147.109	89.063	77.540	75.772	81.187
Atomic volume (m^3/g-atom)	13.1×10^{-6}	23.7×10^{-6}	45.3×10^{-6}	55.9×10^{-6}	70.0×10^{-6}
Atomic radius (Å)	1.225	1.896	2.349	2.48	2.67
Ionic radius (Å)	0.68	0.95[a]	1.33	1.48	1.69
Ionization potential (eV)	5.392	5.139	4.341	4.177	3.894
Electrode potential[b] (V)	−3.024	−2.714	−2.922	−2.99	−2.923
Density (kg/m^3)					
293.15 K	534	968	867	1532	1845
400 K	—	920.87	813.76	1443.61	1787.97
500 K	512.27	897.43	791.34	1397.11	1731.96
600 K	502.17	873.77	768.52	1349.97	1675.96

700 K	492.07	849.93	745.32	1308.57	1619.95
800 K	481.97	825.95	721.78	1266.65	1563.94
900 K	471.87	801.85	697.91	1244.21	1507.94
1000 K	461.77	777.67	673.76	1181.25	1451.93
1100 K	451.67	753.45	649.34	1137.77	1395.93
1200 K	441.57	729.21	624.69	1093.78	1339.92
Specific heat (kJ/kg · K at 298.15 K)	3.549	2.008 (293.15 K)	0.770 (293.15 K)	0.356	0.233
Thermal conductivity (W/m · K)	84.5 (298.15 K)	132.3 (293.15 K)	102.5 (298.15 K)	58.2 (298.15 K)	36.1 (273.2 K)
Vapor pressure [$\log_{10} P$(kPa)]					
300 K		−11.4824	—	—	—
400 K		−6.8657	−4.7911	−3.6998	−3.5058
500 K		−4.1162	−2.5270	−1.5685	−1.5633
600 K		−2.2964	−1.0357	−0.2930	−0.2555
700 K		−1.0054	0.0176	0.5646	0.6557
800 K		−0.0435	0.7997	1.2164	1.3225
900 K		0.6999	1.4026	1.7493	1.8359
1000 K		1.2911	1.8813	2.1913	2.2462
1100 K	−0.2742	1.7720	2.2703	2.5525	2.5801
1200 K	0.3266	2.1705	2.5928	2.8445	2.8526
Electrical resistivity (10^{-8} Ωm at 293 K)	9.28	4.77	7.20	12.84	20.46
Critical temperature (K)	3223 ± 600	2508.7 ± 12.5	2139.9 ± 13.2	2093	2057
Critical pressure (MPa)	68.9 ± 20%	25.64 ± 0.02	16.39 ± 0.02	14.54 ± 0.5	11.73
Critical density (kg/m^3)	120 ± 33	206	—	346	428

[a] 0.95–0.98.
[b] Reduction potential.

superior experimental techniques. When several sets of data were available, they were plotted together as functions of temperature to assess the agreement among them. From these data a temperature-dependent expression which best covers the whole range was selected. The recommended values for the physical properties are listed in Table 2.1. Unless otherwise specified, temperatures are given in Kelvin.

2.1. Electronic Structure and Physical Properties

The properties of alkali metals are directly related to their structure. The electronic structure of alkali metal atoms can be described as an inert-gas or closed-shell configuration with one s-electron outside. The main differences among alkali metal elements arise from the size of the inert-gas shell. The outermost s-electron in the alkali metal atom can be easily released and this accounts for the high chemical reactivity of these elements, which increases down the group. The s-electron also accounts for the metallic nature. As the atomic and ionic sizes increase, the interatomic

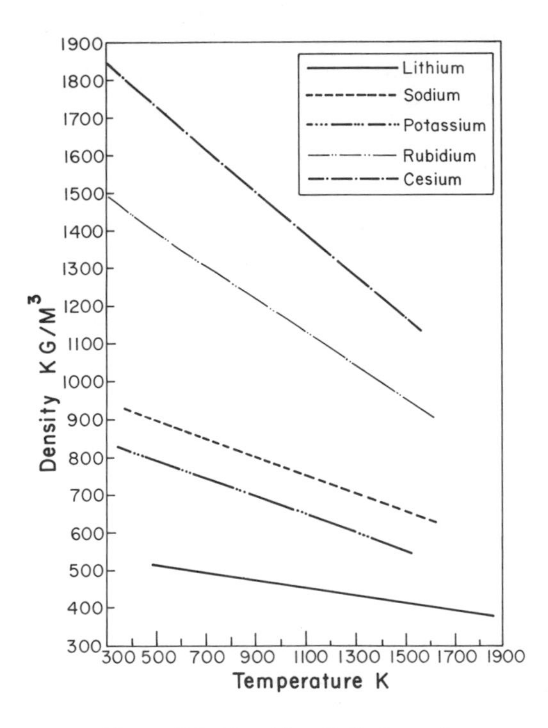

Figure 2.1. Densities of alkali metals.

distances increase and the strength of the metallic bond decreases. This explains the variation of several physical properties as we go down the group.

2.1.1. General Physical Properties

The metals are soft and silver white in color in the pure form. However, a freshly cut metal surface tarnishes quickly even in an atmosphere containing only parts per million levels of oxygen and moisture. The solid metals have a body-centered cubic crystal structure, with the lattice constant increasing from 3.50 Å in the case of lithium to 6.05 Å in the case of cesium. The atomic and ionic radii increase down the group, while the first ionization potential decreases from 5.39 eV in the case of lithium to 3.89 eV in the case of cesium.

The melting point decreases down the group from 180.5°C for lithium to 28.5°C for cesium. Similarly, the boiling point also decreases from 1342°C (Li) to 678.4°C (Cs). Thus, the liquid range is the widest for lithium and the least wide for rubidium and cesium. Melting results in a

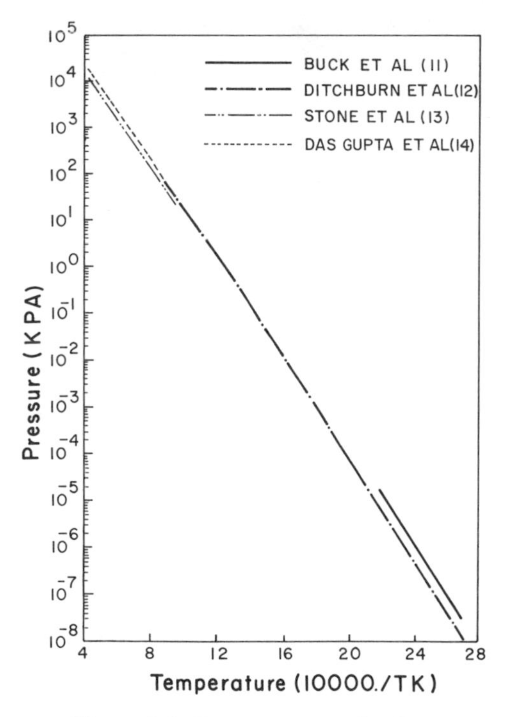

Figure 2.2. Vapor pressure of sodium.

volume increase, the percentage change being higher for the higher alkali metals. The latent heat of fusion of lithium is 3 kJ/g-atom and this decreases down the group to 2.09 kJ/g-atom for cesium. The latent heat of vaporization also shows a similar trend.

Densities are listed in Table 2.1 at several temperatures for easy comparison. Expressions for the temperature dependences of the densities are given in the Appendix. In general, density increases with atomic number except for the reversal of the trend at potassium. The densities of different alkali metals are compared in Fig. 2.1. In Fig. 2.2 the vapor pressure of sodium is plotted as a function of temperature, using data obtained from the available literature.[11-14] The expression given by Das Gupta and Bonilla[14] is recommended. This figure illustrates the method used to select the best values in Vana Varamban's compilation.[1]

2.1.2. Thermophysical Properties

Lithium has the highest specific heat among all elements (3.549 kJ/kg · K at 298 K). Specific heat decreases down the group to a value of 0.233 kJ/kg · K in the case of cesium. Expressions for the temperature dependence of specific heat are given in the Appendix.

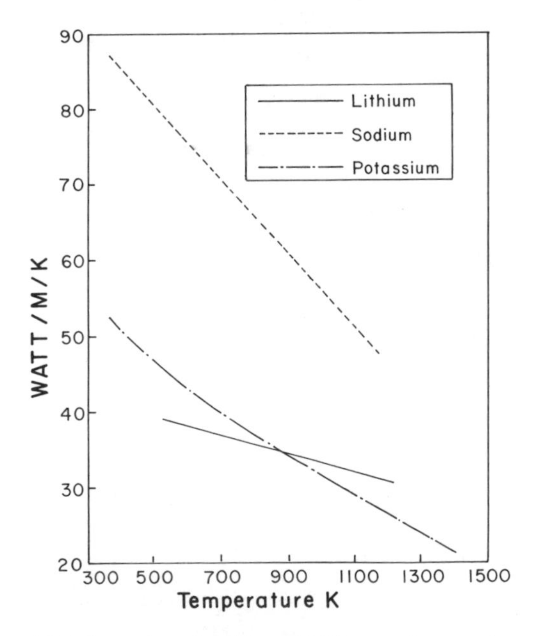

Figure 2.3. Thermal conductivity of lithium, sodium, and potassium.

Sodium has the highest thermal conductivity of the alkali metals (132.3 W/m · K at 298 K), significantly higher than that of lithium (84.8 W/m · K at 298 K). After sodium, the thermal conductivity decreases with increasing atomic number of the alkali metal, reaching a value of 36.1 W/m · K at 273 K for cesium. Expressions for the temperature dependence of thermal conductivity are given in the Appendix. In general, thermal conductivity decreases with increasing temperature, but the temperature dependence is different in the case of lithium. This is seen in Fig. 2.3.

Electrical conductivity shows the same trend as thermal conductivity as one goes down the group. Sodium has the lowest electrical resistivity among alkali metals and cesium the highest.

2.1.3. Vapor Pressure

In the Appendix expressions are given for the vapor pressures of alkali metals in the specified temperature ranges. Table 2.1 also lists vapor pressures at some temperatures. It is seen that vapor pressures are relatively low at the operating temperatures of fast reactors. In Fig. 2.4, we compare the vapor pressures of different alkali metals.

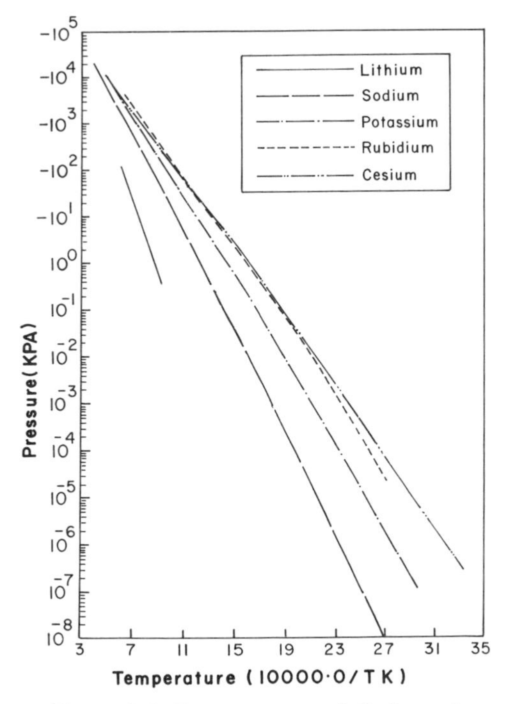

Figure 2.4. Vapor pressure of alkali metals.

2.1.4. Viscosity and Surface Tension

Viscosity and surface tension are important properties which have a bearing on the flow characteristics of the liquid metals. The recommended expressions giving these properties as a function of temperature are listed in the Appendix. Surface tension is especially sensitive to the presence of impurities and the type of cover gas used. This fact has been taken into account in recommending expressions.

Figures 2.5 and 2.6 show the viscosity and surface tension, respectively, of alkali metals at various temperatures. There is a trend of decreasing values as one goes down the group. Both viscosity and surface tension decrease with increasing temperature.

2.1.5. Critical Properties

The critical temperature, critical pressure, and critical density of alkali metals are listed in Table 2.1. The critical temperature decreases from 3223 K in the case of lithium to 2057 K in the case of cesium. Similarly,

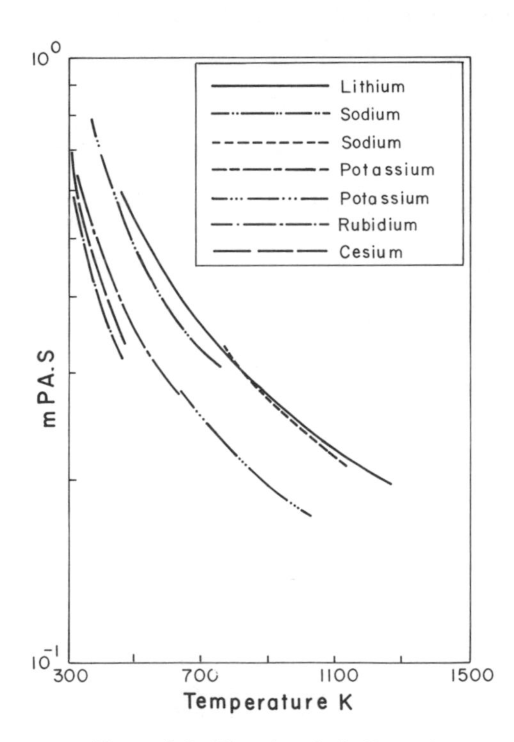

Figure 2.5. Viscosity of alkali metals.

critical pressures range from 68.9 MPa for lithium to 11.7 MPa for cesium. The critical density increases from 120 kg/m^3 (lithium) to 428 kg/m^3 (cesium).

2.2. Nuclear Properties

Sodium and cesium are monoisotopic elements, while lithium and rubidium have two isotopes each and potassium has three. The isotopic compositions as recommended by the International Union of Pure and Applied Chemistry are given in Table 2.2.

In a nuclear reactor (fission or fusion), alkali metal elements are subjected to irradiation with neutrons. The (n, γ) reaction is the most important when the irradiation is with thermal neutrons. In a fast-reactor spectrum other reactions such as (n, p), (n, α), (n, 2n), (n, n′), etc., also become important. Tables 2.3a–e list the cross sections of these processes. In addition to thermal neutron cross sections, fast neutron and 14.5-MeV neutron cross sections are also listed. The fast neutron cross sections refer to the average cross section corresponding to the fast neutron spectrum in a fast breeder reactor.

These nuclear reactions lead to the production of radioactive isotopes. In addition, the fission of heavy elements also leads to the production of

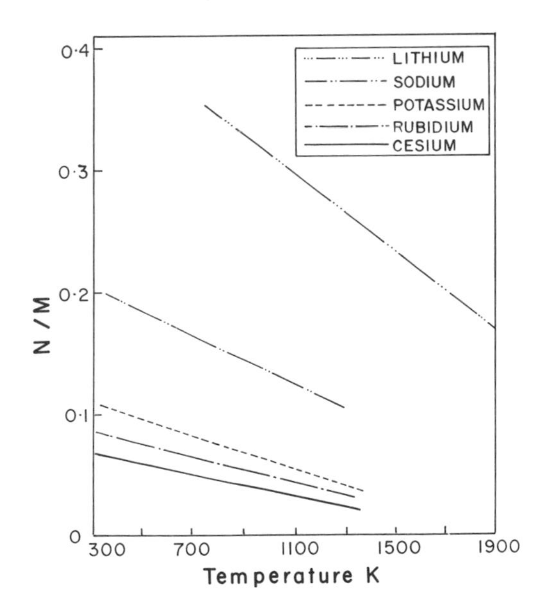

Figure 2.6. Surface tension of alkali metals.

Table 2.2. Isotopic Composition of Alkali Metals[a]

Atomic no.	Element	Mass no.	Evaluated limits of published values (atom %)	Best measurement from a single natural source (atom %)	Representative isotopic composition (atom %)
3	Li	6	7.68–7.30	7.525	7.5
		7	92.70–92.82	92.475	92.5
11	Na	23	—	100	100
19	K	39	—	93.25811	93.2581
		40	—	0.011672	0.0117
		41	—	6.73022	6.7302
37	Rb	85	72.24–72.14	72.1654	72.165
		87	27.86–27.76	27.8346	27.835
55	Cs	133	—	100	100

[a] Data from H. E. Holden and R. L. Martin, *J. Pure and Appl. Chem. 56*, 653–674 (1984).

radioactive isotopes of cesium and rubidium. Some important radioactive isotopes that are likely to be encountered in these processes are also listed in Tables 2.3a–e along with their principal modes of decay. The spectrum-averaged one-group cross sections of a typical fast reactor are given in Table 2.4.

2.3. Chemical Properties

Alkali metals are among the most chemically reactive of all elements. This high chemical reactivity can be attributed to the presence of a single electron in the valence shell, which the atom loses readily to end up as a cation with a stable inert-gas electronic configuration. Chemical reactivity increases down the group, as the valence electron in the heavier alkali metals is more loosely bound because of the increase in shielding from the nuclear charge as the atomic number increases. Thus lithium is the least reactive and cesium the most reactive among the five common alkali metals. The largest difference in reactivity is between lithium and sodium. Lithium, in fact, is the least typical of alkali metals.

These differences are clearly observed in the reactions of alkali metals with common elements and compounds. Thus lithium does not react with oxygen at room temperature, but sodium and potassium react fairly rapidly. Rubidium and cesium burn in oxygen. Because of their great affinity for oxygen, alkali metals can be used to reduce many metal oxides to produce metals. For example, sodium reduces chromium oxide and manganese oxide to the metals. In fact, the Woehler process for producing aluminum used potassium metal as a reducing agent.

All alkali metals react with water. Lithium reacts only slowly, but the vigor of the reaction increases down the group, becoming quite violent in the case of the heavy alkali metals. Sodium reacts even with ice. The reaction between sodium and water can be explosive if the interface between the two phases is large. As this reaction is important in sodium-heated steam generators, it has been extensively studied. In the presence of excess of water, the reaction takes the familiar course:

$$2Na + 2H_2O \rightarrow 2NaOH + H_2\uparrow \quad (1)$$

In the case of a leak in a sodium-heated steam generator, it is high-pressure steam or water that leaks into the sodium. Thus there is excess sodium and hence the reaction products in the above primary reaction further react with excess sodium:

$$Na + NaOH \rightarrow Na_2O + NaH \quad (2)$$

$$Na + \tfrac{1}{2}H_2 \rightarrow NaH \quad (3)$$

Monitoring oxygen or hydrogen in a flowing sodium system can, therefore, indicate leakage of water into the system.

Hydrogen reacts with alkali metals to form the hydrides, MH. The reaction takes place with the liquid metal, but at relatively low temperatures the hydride film which forms on the liquid surface retards further

Table 2.3a. Neutron Cross Sections of Lithium Isotopes

Energy of the neutrons	Target nuclide	Abundance	Reactions and product	Half-life of the product	Cross section	
					σ	I
Slow	Li-6	7.5	(n, α) H-3	12.346 a	940 ± 4 b	425 b
			(n, γ) Li-7	Stable	50 mb	21 mb
	Li-7	92.5	(n, γ) Li-8	844 ms	37 ± 4 mb	16 mb
	Li (natural)		(n, γ)	—	70.7 ± 7 b	32 b
Fast	Li-6	7.5	(n, p) He-6	802 ms	39 mb	—
			(n, 2n) α, p	Stable	2.1 mb	—
	Li-7	92.5	(n, p) He-7	—	<0.1 μb	—
			(n, 2n) Li-6	Stable	370 mb	—
14.5 MeV	Li-6	7.5	(n, p) He-6	802 ms	8.3 ± 2 mb	—

Table 2.3b. Neutron Cross Sections of Sodium Isotopes

Energy of neutrons	Target nuclide	Abundance/ half-life	Reactions and product	Half-life of the product	Cross section		Major gamma lines of the product (keV) (abundance in parentheses)
					σ	I	
Slow	Na-23	100	(n, γ) Na-24 m	20.1 ms	400 ± 30 mb	—	472 (100)
			(n, γ) Na-24 g (cum)	15.02 h	530 ± 5 mb	340 ± 10 mb[a]	1368.6 (100), 2754.1 (100)
	Na-22 (artificial)	2.601 a	(n, γ) Na-23	Stable	$29{,}000 \pm 1000$ b	$160{,}000 \pm 21{,}000$ b	—
Fast	Na-23	100	(n, p) Ne-23	37.6 s	1.5 ± 0.6 mb[b]	—	439 (100)
			(n, α) F-20	11.0 s	765 ± 170 μb[c]	—	1633.1 (100)
			(n, 2n) Na-22	2.601 a	2.2 ± 0.2 μb[d]	—	511 (181), 1274.5 (90)
14.5 MeV	Na-23	100	(n, p) Ne-23	37.6 s	43 ± 6 mb	—	439 (100)
			(n, α) F-20	11.0 s	150 ± 15 mb	—	1633.1 (100)
			(n, 2n) Na-22	2.601 a	44 ± 3 mb	—	511 (181), 1274.5 (90)

[a] Other value, 311 ± 10 mb.
[b] Other value, 310 μb.
[c] Other value, 490 μb.
[d] Other value, 8 μb.

Table 2.3c. Neutron Cross Sections of Potassium Isotopes

Energy of neutrons	Target nuclide	Abundance/ half-life	Reactions and product	Half-life of the product	Cross section σ	Cross section I	Major gamma lines of the product (keV) (abundance in parentheses)
Slow	K (natural)		(n, γ)		2.1 ± 1 b	1.0 ± 0.1 b	
	K-39	93.3	(n, γ) K-40	1.28×10^9 a	1.96 ± 0.15 b	—	
			(n, α) Cl-36	3.01×10^5 a	4.3 ± 0.5 mb	—	
	K-40	0.0118	(n, γ) K-41	Stable	30 ± 8 b[a]	31 b	
			(n, p) Ar-40	Stable	4.4 ± 0.3 b	—	
			(n, α) Cl-37	Stable	390 ± 30 mb	—	
	K-41	6.7	(n, γ) K-42	12.36 h	1.46 ± 0.03 b	1.42 ± 0.06 b[b]	1524.7 (17.9)
Fast	K-39	93.3	(n, p) Ar-39	269 a	20 mb	—	
			(n, α) Cl-36	3.01×10^5 a	13 mb[c]	—	
			(n, 2n) K-38 m + g[f]	929 ms + 7.63 m	7 μb	—	
	K-40	0.0118	(n, α) Cl-37	Stable	31 mb	—	
			(n, 2n) K-39	Stable	1.6 mb	—	
	K-41	6.7	(n, p) Ar-41	1.83 h	2.1 ± 0.2 mb[d]	—	1293.6 (99.2)
			(n, α) Cl-38 g (cum)	37.2 m	760 ± 50 μb[e]	—	1642.4 (32.8), 2167.5 (44)
			(n, 2n) K-40	1.28×10^9 a	160 μb	—	
14.5 MeV	K-39	93.3	(n, p) Ar-39	269 a	354 ± 45 mb	—	
			(n, α) Cl-36	3.01×10^5 a	84 ± 12 mb (14.1)	—	
			(n, 2n) K-38 m[f]	929 ms	800 ± 200 μb (14.9)[g]	—	511 (200)
			(n, 2n) K-38 g	7.63 m	$3.5 \pm$ mb[g]	—	511 (200), 2166.8 (100)
	K-41	6.7	(n, p) Ar-41	1.83 h	49 ± 5 mb	—	1293.6 (99.2)
			(n, α) Cl-38 g (cum)	37.2 m	39 ± 8 mb	—	1642.4 (32.8), 2167.5 (55)
			(n, 2n) K-40 m	0.294 μs	36 ± 16 mb	—	

[a] Other value, 70 b.
[b] Other values, 1.28 b, 0.96 b, 0.96 b, 1.35 b.
[c] Other value, 8 ± 0.3 mb.
[d] Other value, 1.1 mb.
[e] Other value, 2.6 mb.
[f] No IT.
[g] Other value for K-39 (n, 2n) K.38 m + g: 4.36 ± 0.28 mb (14.7 MeV).

Table 2.3d. Neutron Cross Sections of Rubidium Isotopes

Energy of neutrons	Target nuclide	Abundance/ half-life	Reactions and product	Half-life of the product	Cross section		Major gamma lines of the product nuclide (keV) (abundance in parentheses)
					σ	I	
Slow	Rb-85	72.17	(n, γ) Rb-86 m[b]	1.018 m	50 ± 5 mb	1.16 ± 0.03 b	555.8 (98.2)
			(n, γ) Rb-86 g	18.65 d	410 ± 20 mb	3.65 ± 0.2 b	
			(n, γ) Rb-86 g (cum)	18.65 d	460 ± 20 mb	7.5 ± 0.5 b	1076.6 (8.76)
	Rb-87	27.83	(n, γ) Rb-88	17.7 m	120 ± 30 mb	2.0 ± 0.5 b	898 (14.5), 1836.1 (21.4), 2776.4 (21)
	Rb (natural)		(n, γ)		370 ± 30 mb[a]	6.0 ± 0.5 b	
Fast	Rb-85	72.17	(n, p) Kr-85 m + g[d]	4.48 h & 10.73 a	850 μb		
			(n, α) Br-82 m + g (cum)[e]	6.1 m & 35.4 h	11 μb		554.3 (72), 619.1 (39), 698.4 (28), 776.5 (83.2), 1044 (28), 1317.4 (27)
			(n, 2n) Rb-84 m + g (cum)[b]	20.4 m & 34 d	200 ± 100 μb		511 (42), 881.5 (75.3)
	Rb-87	27.83	(n, p) Kr-87	76 m	7 μb		402.7 (48), 845.6 (7.3), 2554.5 (8.7)
			(n, α) Br-84 m + g[f]	6 m & 31.8 m	0.4 μb		
			(n, 2n) Rb-86 m + g (cum)[b]	1.018 m & 18.65 d	350 μb		1076.6 (8.76)

14.5 MeV	Rb-85	72.17	(n, p) Kr-85 m[d]	4.48 h	4.1 ± 0.4 mb	151 (76.1), 304.5 (13.5)
			(n, α) Br-82	35.4 h	7 ± 2 mb	554.3 (72), 619.1 (39), 698.4 (28), 776.5 (83.2), 1044 (28), 1317.4 (27)
			(n, 2n) Rb-84 m[b]	20.4 m	$(560 \pm 220$ mb)[g]	
			(n, 2n) Rb-84 m + g (cum)	34 d	1.411 ± 0.086 b[h]	511 (42), 881.5 (75.3)
	Rb-87	27.83	(n, p) Kr-87	76 m	$(4.9 \pm 0.5$ mb)[i]	402.7 (48.3), 845.6 (7.3), 2554.5 (8.7)
			(n, α) Br-84 m[f]	6 m	1.9 ± 0.4 mb	424 (100), 881.5 (98), 1462.8 (97)
			(n, α) Br-84 g	31.8 m	1.8 ± 0.2 mb	881.6 (41), 802.2 (6), 1015.9 (6.2), 1897.6 (14.7), 2484.1 (6.7), 3927 (6.8)
			(n, 2n) Rb-86 m[b]	1.018 m	$(760 \pm 250$ mb)[g]	
			(n, 2n) Rb-86 m + g (cum)	18.65 d	1.817 ± 0.343 b[k]	1076.6 (8.76)

[a] Other value, 700 ± 7 mb.
[b] 100% IT.
[d] 20.4% IT.
[e] 99% IT.
[f] No IT.
[g] Unweighted mean of several unrecommended values.
[h] Other value, 1.3 ± 0.1 b; unweighted mean of several unrecommended values.
[i] Other value, $10 + 2$ mb. Both values are not recommended.
[k] Other value, 1.4 ± 0.1 b; unweighted mean of several unrecommended values.

Table 2.3e. Neutron Cross Sections of Cesium Isotopes

Energy of neutrons	Target nuclide	Abundance/ half-life	Reactions and product	Half-life of the product	Cross section σ	Cross section I	Major gamma lines of the product (keV) (abundance in parentheses)
Slow	Cs-133	100	(n, γ) Cs-134 m[b]	2.9 h	2.5 ± 0.2 b	29.2 ± 6.2 b	30.9 (18), 127.4 (14)
			(n, γ) Cs-134 g	2.06 a	27 ± 1.5 b	359 ± 90 b[c]	563.3 (8.4), 569.3 (15.4), 604.7 (97.6), 795.8 (85.4), 801.8 (8.73)
			(n, γ) Cs-134 g (cum)		29.0 ± 1.5 b	415 ± 15 b[d]	
	Cs-134	2.06 a	(n, γ) Cs-135	23×10^6 a	140 ± 12 b		
	Cs-135	23×10^6 a	(n, γ) Cs-136	13.00 d	8.7 ± 0.5 b	62 b	
	Cs-136	13.00 d	(n, γ) Cs-137[e]	30.1 a	1.3 b	76 b[f]	
	Cs-137	30.1 a	(n, γ) Cs-138	32.2 m	110 ± 33 b	414 mb	
Fast	Cs-133	100	(n, p) Xe-133 m + g[g]	2.23 d & 5.29 d	81 μb		
			(n, α) I-130 m + g[h]	8.9 m & 12.4 h	3.3 ± 0.8 μb[i]		418 (34.2), 536.1 (99.5) 668.5 (96.7), 739.5 (82.7), 1157.5 (11)
			(n, 2n) Cs-132	6.47 d	1.2 mb		29.7 (60), 667.7 (100)
Fast	Artificial nuclides						
	Cs-135	23×10^6 a	(n, p) Xe-135 m + g[k]	15.3 m & 9.17 h	25 μb		
			(n, α) I-132	2.285 h	0.7 μb		
			(n, 2n) Cs-134 m + g[b]	2.9 h & 2.06 a	1.5 mb		
	Cs-137	30.1 a	(n, p) Xe-137	3.84 m	0.2 μb		
			(n, α) I-134 m + g[l]	3.6 m & 52.6 m	0.3 μb		
			(n, 2n) Cs-136	13.00 d	2.5 mb		

14.5 MeV	Cs-133	100	(n, p) Xe-133 m[b]	2.23 d	4.8 ± 0.75 mb	29.7 (46.5), 233.2 (8.85)
			(n, p) Xe-133 m + g	5.29 d	5.7 ± 2.35 mb	30.9 (37.5), 81 (36.5)
			(n, α) I-130	12.4 h	1.9 ± 0.3 mb	418 (34.2), 536.1 (99.5), 668.5 (96.7), 739.5 (82.7), 1157.5 (11)
			(n, 2n) Cs-132	6.47 d	1.52 ± 0.11 b	29.7 (60), 667.7 (100)

[b] ~100% IT.
[c] Other value, 400 b.
[d] Other values, 450 b, 407 ± 60 b.
[e] Daughter Ba-137 m.
[f] Other value, 15.55 b.
[g] 100% IT. Decay chain.
[h] 83% IT.
[i] IT included? Other value, 2 μb.
[k] Decay chain.
[l] ~98% IT.

Explanations for Table 2.3

Tables 2.3a–e provide nuclear data for the alkali metals. The various parameters used in these tables are defined here.
Column 1 gives the energy of the neutrons in terms of the following classification:

1. Slow (reactor) neutrons (includes both thermal and epithermal neutrons). The main reaction is of the (n, γ) type.
2. Fast (reactor) neutrons (energies of ~100 keV and above).
3. 14.5 MeV neutrons from neutron generators.

In column 2, the target nuclide is identified; in column 3 its abundance/half-life is given. The usual notations for the half-life of a nuclide are used: a = year; d = day; h = hour; m = minute; s = second. Abundance is given in percent.

In column 4, the type of reaction and the reaction product are indicated; column 5 gives the half-life of the product. The cross-sections are given in column 6. For slow (reactor) neutrons σ_{thermal} is given; I is the resonance integral. The σ values quoted for fast (reactor) neutrons are average cross sections for a fast reactor spectrum. For 14.5-MeV neutrons the σ values corresponding to the Texas convention are given. Details can be found in Ref. 10. The cross section values are expressed in barns (b), millibarns (mb), or microbarns (μb); 1 barn $= 10^{-24}$ cm^2.

In column 6, the γ-lines of the activation products are listed. Energies are in keV; absolute intensities are in percent (photons per 100 decay events).

Table 2.4. One-Group Cross Sections for Alkali Metal Isotopes (barns)

	^{6}Li	^{7}Li	^{23}Na	K	^{85}Rb	^{133}Cs
(n, n)	0.15847E+01	0.15501E+01	0.57954E+01	0.23876E+01	0.72887E+01	0.68855E+01
(n, γ)	0.30252E−04	0.35434E−04	0.34071E−02	0.12131E−01	0.22269E+00	0.48040E+00
(n, p)	0.35617E−03	—	0.14171E−03	0.92323E−02	—	0.25756E−04
(n, d)	—	0.72406E−06	—	—	—	—
(n, α)	0.11337E+01	—	0.73060E−04	0.20926E−02	—	0.37656E−06
(n, 2n)	—	0.37932E−05	0.19544E−06	0.20750E−07	—	0.25459E−03
(n, n)	0.15185E−01	0.30757E−01	0.99715E−01	0.10571E−01	—	0.51854E−00
Total	0.27340E+01	0.15809E+01	0.58987E+01	0.24216E+01	0.75114E+01	0.78847E+01

reaction. At higher temperatures the reaction proceeds rapidly. Hydrogen absorption by sodium is quite rapid above 300°C.

Nitrogen reacts with lithium to form Li_3N. Hence nitrogen cannot be used as a cover gas for lithium. Under ordinary conditions, the other alkali metals do not react with nitrogen.

All alkali metals react strongly with halogens, the vigor of the reaction increasing with the atomic number of the alkali metal. While lithium reacts readily with halogens, emitting light, sodium ignites in fluorine. The reaction of the heavier alkali metals is violent with all halogens, leading to ignition and/or detonation. Because of their high affinity for halogens, alkali metals react with compounds containing halogens. They react vigorously with most inorganic halides, reducing them to metals. In fact, such reductions with sodium are commercially important as they can be used in the manufacture of metals such as titanium and zirconium.

Alkali metals do not react with carbon except at high temperatures. Lithium and sodium form acetylides (M_2C_2) at high temperatures. The heavier metals are not known to form carbides, but carbon dissolves in them to some extent. The chemistry of carbon and nitrogen in liquid alkali metals is discussed in Chapter 6. All alkali metals react with carbon dioxide. Under controlled conditions, sodium reacts with CO_2 to form sodium formate and sodium oxalate but at red heat, sodium carbonate and carbon are formed. Molten sodium detonates in contact with solid CO_2 and hence CO_2 cannot be used to extinguish sodium fires. Lithium reacts with CO_2 only at high temperatures, but the heavier alkali metals react progressively more rapidly than sodium. Reaction of CO_2 with potassium results in the formation of K_2O and potassium oxalate. Potassium, rubidium, and cesium form lamellar compounds with graphite, in which the alkali metals are believed to be loosely bonded between parallel planes of carbon atoms.

Ammonia reacts with alkali metals to form amides, MNH_2. The reaction is slow in the case of lithium and sodium, but increases in rate from potassium to cesium. The characteristic deep blue color that results when sodium dissolves in liquid ammonia at low temperatures is well known. This solution, whose blue color is due to solvated electrons, is a strong reducing agent. In the presence of metal catalysts this solution decomposes to yield sodamide, $NaNH_2$. At high temperatures ammonia gas reacts with liquid sodium to form $NaNH_2$. Reaction of sodium, ammonia, and carbon results in sodium cyanide.

2.4. Alkali Metals in Organic Reactions

The reactivity and reducing properties of alkali metals are made use of in organic syntheses. Thus triple bonds can be selectively reduced to double bonds and phenolic esters are reduced to benzenes using potassium/sodium in liquid ammonia. In Birch reduction, aromatic rings are reduced by sodium, potassium, or lithium in ammonia to nonconjugated cyclohexadienes. Simple straight-chain carboxylic acids are reduced to aldehydes by treatment with lithium in liquid ammonia, followed by the hydrolysis of the resulting imine. The same reagent reduces aryl aldehydes to methyl benzenes. The affinity of alkali metals for halogens is taken advantage of in dehalogenation reactions using sodium in ammonia. The same reactive property finds application in the production of tetraethyl lead by the action of ethyl chloride on sodium–lead alloy

$$4C_2H_5Cl + Na_4Pb \rightarrow (C_2H_5)_4Pb + 4NaCl \qquad (4)$$

Alkali metals also figure in synthesis through coupling reactions. In the Wurtz reaction, coupling of alkyl halides takes place through the agency of sodium

$$2RX + 2Na \rightarrow RR + 2NaX \qquad (5)$$

Similarly, small rings can be closed through the reaction of halogen derivatives with sodium

$$\mathrm{Br{-}\Diamond{-}Cl + 2Na \longrightarrow \Diamond + NaCl + NaBr} \qquad (6)$$

In the Wurtz–Fittig reaction, a mixture of alkyl and aryl halides is treated with sodium to give an alkylated aromatic compound. Potassium in ammonia is used for the arylation of aliphatic compounds such as ketones. Alkylation of an α,β-unsaturated ketone is brought about through the agency of lithium in ammonia

$$\begin{array}{c} | \quad\ \ | \\ -\mathrm{C}=\mathrm{C}-\mathrm{C}- \\ | \qquad\quad \| \\ \mathrm{H} \qquad\ \ \mathrm{O} \end{array} \xrightarrow{\mathrm{Li/NH_3}} \begin{array}{c} | \quad\ \ | \\ -\mathrm{C}-\mathrm{C}-\mathrm{C}- \\ | \quad\ \ | \quad\ \ \| \\ \ \ \mathrm{R} \quad \mathrm{O} \end{array} \qquad (7)$$

Pinacols can be synthesized by the reduction of aldehydes and ketones with active metals such as sodium

$$R-\underset{\underset{O}{\|}}{C}-R \xrightarrow{Na} R-\underset{OH}{\overset{R}{\underset{|}{\overset{|}{C}}}}-\underset{OH}{\overset{R}{\underset{|}{\overset{|}{C}}}}-R \tag{8}$$

Alkali metals react with alcohols and phenols to give alkoxides and phenoxides, respectively, which, in turn, are useful intermediates in organic synthesis. For example, in the Williamson synthesis, an alkoxide reacts with an alkyl chloride to yield an ether

$$RONa + RCl \rightarrow ROR + NaCl \tag{9}$$

In the Claisen condensation, esters with α-protons react in the presence of a base to yield β-keto esters

$$2CH_3COOC_2H_5 \xrightarrow{C_2H_5ONa} CH_3\cdot CO\cdot CH_2\cdot COOC_2H_5 + C_2H_5OH \tag{10}$$

Salicylic acid is synthesized by the Kolbe method by heating dry sodium phenoxide with carbon dioxide at 125–140°C under six to seven atmospheres pressure.

Appendix

Table 2.A1. The Temperature-Dependent Expressions for Various Physical Properties

Element	Temperature range (K)	Expression
Density (kg/m^3)		
Lithium	473–1873	$\rho_T = 515 - 0.101(T - 473)$
Sodium		
Solid	273–370.98	$\rho_T = 972.5 - 20.11 \times 10^{-2}(T - 273.15) - 1.5 \times 10^{-4}(T - 273.15)^2$
Liquid	370.98–1644.24	$\rho_T = 1011.8 - 0.22054T - 1.9226 \times 10^{-5}T^2 + 5.637 \times 10^{-9}T^3$
	Above 1644	$\rho_T = 214.1[1 + 2.3709(1 - T/2509.46)^{0.31645} + 2.8467 \times 10^{-7}(2509.46 - T)^2]$
Potassium		
Solid	4.15–336.35	$\rho_T = 864 - 2.4162 \times 10^{-1}(T - 273.15)$
Liquid	336.35–1523.15	$\rho_T = 841.5 - 2.172 \times 10^{-1}(T - 273.15) - 2.70 \times 10^{-5}(T - 273.15)^2 + 4.77 \times 10^{-9}(T - 273.15)^3$
Rubidium		
Liquid	312.47–523.15 (m.p.)	$\rho_T = 1484 - 4.65 \times 10^{-1}(T - 313.15)$
Liquid	573.15–1623.15	$\rho_T = 1361 - 4.1 \times 10^{-1}(T - 573.15) - 2.6 \times 10^{-5}(T - 573.15)^2$
Cesium		
Liquid	Below 1573	$\rho_T = 1747 - 5.6 \times 10^{-1}(T - 473.15) - 6.2 \times 10^{-5}(T - 473.15)^2$

Specific heat (kJ/kg)		
Lithium	453.69–694.95	$C_p = 1.083 - 2.00 \times 10^{-4}(T - 273.15)$
	694.95–1178.45	$C_p = 1.006 - 1.73 \times 10^{-5}(T - 273.15)$
Sodium	273.15–370.98	$C_p = 1.1987 + 6.4894 \times 10^{-4}(T - 273.15) + 1.053 \times 10^{-5}(T - 273.15)^2$
	370.98–1173.15	$C_p = 1.4361 - 5.8024 \times 10^{-4}(T - 273.15) + 4.621 \times 10^{-7}(T - 273.15)^2$
Potassium	100–336.35	$C_p = 0.5376 + 7.998 \times 10^{-4}T$
	336.35–1423	$C_p = 0.8385 - 3.6723 \times 10^{-4}(T - 273.15) + 4.58985 \times 10^{-7}(T - 273.15)^2$
Rubidium		$C_p = 0.38317 - 1.09495 \times 10^{-4}T + 1.020896 \times 10^{-7}T^2$
Thermal conductivity (W/m · K)		
Lithium	523–1223	$K = 42.3 + 12.3 \times 10^{-3}(T - 273.15)$
Sodium	370.98–1173.15	$K = 91.8 - 4.9 \times 10^{-2}(T - 273.15)$
Potassium	2.3–336.35	$K = 125.938 - 6.02496 \times 10^{-2}T$
	373–1423	$K = 43.8 - 2.22 \times 10^{-2}(T - 273.15) + (39.5 \times 10^2)/T$
Vapor pressure (kPa)		
Lithium	1073–1673	$\log_{10} P = 1.054165 - (839.2851/T) - (7.1748 \times 10^{-2} \log_{10} T)$
Sodium	300–2500	$\log_{10} P = 10.182516 - (5693.8776/T) - (1.0948 \log_{10} T) + 8.5874946 \times 10^{-5}T$
Potassium	336.35–2200	$\log_{10} P = 10.146317 - (4778.5/T) - (1.1032 \log_{10} T) - 0.4093 \times 10^{-3}T + 0.29444 \times 10^{-6}T^2 - 0.0621 \times 10^{-9}T^3$
Rubidium	370–1644	$\log_{10} P = -46.4339996 + 0.25791711T - 6.3849322 \times 10^{-4}T^2 + 9.0126748 \times 10^{-7}T^3 - 7.582497 \times 10^{-10}T^4 + 3.7638599 \times 10^{-13}T^5 - 1.0175407 \times 10^{-16}T^6 + 1.1551688 \times 10^{-20}T^7$

Table continued

Table 2.A1. *(continued)*

Element	Temperature range (K)	Expression
Cesium	301.67–1873.15	$\log_{10} P = -2.7128664 \times 10^{1} + 1.1592692 \times 10^{-1}T - 2.1216135 \times 10^{-4}T^2 + 2.2182996 \times 10^{-7}T^3 - 1.3339542 \times 10^{-10}T^4 + 4.2871196 \times 10^{-14}T^5 - 5.6964257 \times 10^{-18}T^6$
Viscosity (mPa · s)[a]		
Lithium	Below 1273	$\log_{10} \eta(\mathrm{Pa \cdot s}) = -1.4936 - 0.7368 \log_{10} T + 109.95/T$
Sodium	Below 773	$\eta = (0.1235 \pm 0.0018)(\rho/1000)^{1/3} \exp \frac{(697 \pm 9)}{1000T}$
	Above 773	$\eta = (0.0851 \pm 0.0013)(\rho/1000)^{1/3} \exp \frac{(1040 \pm 19)}{1000T}$
Potassium	Below 653	$\eta = (0.1131 \pm 0.0060)(\rho/1000)^{1/3} \exp \frac{(680 \pm 40)}{1000T}$
	Above 653	$\eta = (0.0799 \pm 0.0032)(\rho/1000)^{1/3} \exp \frac{(978 \pm 59)}{1000T}$
Rubidium	312.47–473	$\ln \eta = -2.401 + 582/T$
Cesium	301.67–473	$\ln \eta = -2.322 + 584/T$
Surface tension (N/m)		
Lithium	773.15–1873.15	$\sigma = 1.6 \times 10^{-4}(3550 - T) - 95 \times 10^{-3}$
Sodium	370.98–1300	$\sigma = 206.7 \times 10^{-3} - 1.0 \times 10^{-4}(T - 273.15)$
Potassium	336.35–1373	$\sigma = 107.1 \times 10^{-3} - 6.9 \times 10^{-5}(T - 336.35)$
Rubidium	312.47–1373	$\sigma = 85.7 \times 10^{-3} - 5.3 \times 10^{-5}(T - 311)$
Cesium	301.67–1373	$\sigma = 68.8 \times 10^{-3} - 4.5 \times 10^{-5}(T - 301)$

[a] ρ in viscosity expressions is density in kg/m^3.

References

1. S. Vana Varamban, Physical Properties of Alkali Metals, RRC Report, Reactor Research Centre, Kalpakkam, India, 1985.
2. R. N. Lyon (Ed.), *Liquid Metals Handbook*, 2nd Ed., U.S. Govt. Printing Office, Washington D.C., 1952.
3. R. C. Weast (Ed.), *Handbook of Chemistry and Physics*, 68th Ed., Chemical Rubber Co. Press Inc., West Palm Beach, Florida, 1982/83.
4. O. J. Foust (Ed.), *Sodium-NaK Engineering Handbook*, Vol. 1, Sodium Chemistry and Physical Properties, Gordon and Breach, New York, 1972.
5. F. M. Perel'Man, *Rubidium and Cesium*, Macmillan, New York, 1965.
6. G. W. Thomson and E. Garalis, in: *Sodium—Its Manufacture, Properties and Uses* (M. Sittig, Ed.), Reinhold Publishing Corporation, New York, 1956.
7. A. Padilla, Jr., High Temperature Thermodynamic Properties of Sodium, HEDL-TME 77-27, UC-796, Hanford Engineering Development Laboratory, Richland, Wash., 1978.
8. J. K. Fink and L. Leibowitz, Thermophysical Properties of Sodium, ANL-CEN-RSD-79-1, Argonne National Laboratory, Argonne, Ill., 1979.
9. R. T. Sanderson, *Chemical Periodicity*, Reinhold Publishing Corporation, New York, 1960.
10. G. Erdtmann, *Neutron Activation Table*, Kernchemie in Einzeldarstellungen, Vol. 6, Verlag Chemie, Weinheim, New York, 1976.
11. U. Buck and H. Pauly, *Z. Phys. Chem. 44*, 345–352 (1965).
12. R. W. Ditchburn and J. C. Gilmour, *Rev. Mod. Phys. 13*, 310–327 (1941).
13. J. P. Stone, C. T. Ewing, J. R. Spann, E. W. Steivikuller, D. D. Williams, and R. R. Miller, *J. Chem. Eng. Data 11*(3), 309–314 (1966).
14. S. Das Gupta and C. F. Bonilla, cited in Ref. 8.

3

Production of Alkali Metals

The chemical reactivity and the physical properties of the alkali metals govern the choice of the metallurgical processes for their industrial production. The large free energy of formation of the alkali salts and oxides makes it necessary to apply strong reducing agents or the electrolysis of their molten hydroxides or salts to convert them into the metallic state.

The production of sodium is much higher than that of any other alkali metal. Therefore, the processes to produce sodium are of great economic importance. The total capacity for sodium metal production in the Western countries of the world is about 250,000 t/a, two-thirds of which is in the USA. The other alkali metals do not play a similar role. Their production does not exceed some hundred tons per year, for instance, 130 tons of potassium in 1971. Information concerning the capacities for lithium production are not available, but only a very few companies are involved in the production of this metal. However, the technological applications of lithium seem to be increasing. This tendency may cause an increase in its production. Rubidium and cesium are rare and expensive and, therefore, their production is limited.

Lithium metal is produced on an industrial scale by means of the electrolysis of molten lithium chloride, the melting point of which is lowered by the addition of potassium chloride. The eutectic of these salts containing 42.6 wt% lithium chloride has a melting point of 325°C (598 K). The concentration of lithium chloride in the process mixture is higher than in the eutectic mixture (45–55 wt%), and hence the temperature of the molten salts during the electrolysis is kept at 400 to 460°C (673–733 K). The decomposition potential of lithium chloride is 3.68 V; however, 6 to 6.5 V are necessary for the electrochemical reduction of the molten lithium salt.

The principle of the electrochemical cell is the same as generally applied for alkali metal production. The electrolytical process can be continuously performed, since the salt mixture can be introduced into the cell and the molten lithium can be taken out without any disturbance of the process.

The material of the container and the cathode is cast steel, while the anode is made of graphite. A diaphragm of steel mesh divides the anode area, in which the chlorine is generated, from the cathode section. Figure 3.1 gives a description of the principle of the cell with a diaphragm. If the distance between the anode and cathode is chosen to be larger, the diaphragm is not necessary. However, such cells have a higher energy consumption. The molten lithium metal is collected on the top of the molten electrolyte and can be scooped out of the cell.

The liquid metal is poured into containers to form bars at temperatures just above the melting point. Before the molten lithium is cast, it is aged for some time in order to precipitate nonmetallic impurities. The lithium metal freezes in cylinders or in molds of steel, in sizes of 0.1, 1, 4, and 25 kg. The bars or cylinders of the metal are protected against contamination from the atmosphere. This is normally done by covering the bars with petrol ether or by replacing the atmosphere by pure argon cover gas. Nitrogen cannot be applied as protective gas, since it reacts to form lithium nitride, which is soluble in the metal even at low temperatures. The soft metal can be mechanically formed into wires, hoops, or foils. All these deformation processes have to be performed under protection from the atmosphere.

The lithium chloride used in the electrolysis is prepared from lithium ores. The ores are first dissolved in sulfuric acid at a temperature between 250 and 400°C. The solution of lithium sulfate is treated with soda and lime in order to precipitate the alkaline earths, aluminum, and iron. Finally, lithium carbonate is precipitated at a temperature close to the boiling point by addition of a concentrated solution of soda. This procedure has a yield of about 90% with respect to the content of lithium and a relatively acceptable energy balance.

The annual production of lithium ores ranges from 200,000 to 300,000 tons. However, only a small proportion of this amount is used for the production of the metal. Lithium in the metallic state is mainly used as an alloying element in order to improve the properties of several aluminum-based alloys. Its high getter capacity and its low density are the properties responsible for its usefulness in alloying techniques. A more recent application of the metal is its use in electrochemical batteries or as blanket fluid in fusion reactors.

Several methods to produce pure lithium (and other alkali metals) on a laboratory scale have been published. Thus, alkali chromates, molyb-

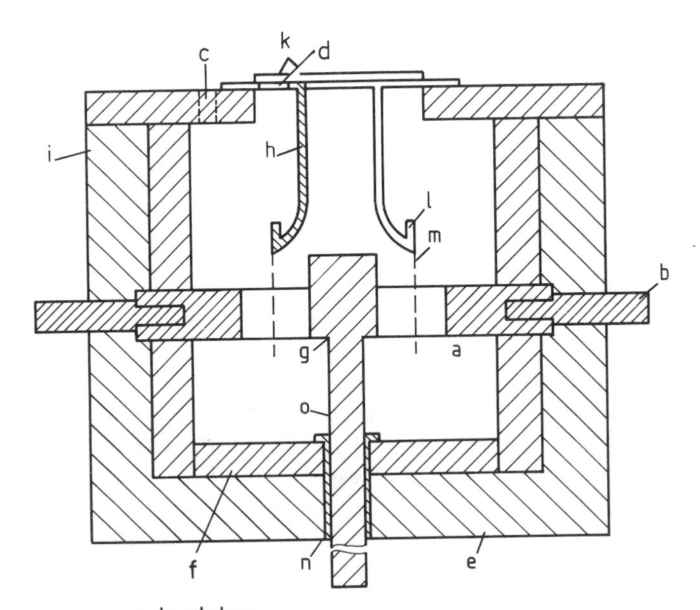

a anode plates
b graphite anodes
c chlorine outlet
d opening for introduction of LiCl
e insulation
f fire-resistant plates
h collector for lithium metal (of cast iron)
g head of the steel cathode
i container of steel
k steel plates for closure of the cell
l holder for the diaphragm
m wire net as diaphragm
n steel ring for support of the cathode
o shaft of the cathode

Figure 3.1. Modified version of the Downs cell used for the production of lithium metal by means of electrolysis of molten chlorides.

dates, or tungstates are chemically reduced by zirconium powder in high vacuum at a temperature of 450 to 600°C according to equation (1).[1]

$$2Li_2CrO_4 + Zr \rightleftarrows 4Li + Zr(CrO_4)_2 \quad (1)$$

Very pure alkali metals are the products of the decomposition of alkali azides in high vacuum at relatively low temperatures. The reaction proceeds even in a laboratory apparatus made of glass. Lithium, however, reacts more or less violently with glass. The production of this metal cannot, therefore, be carried out in glass vessels. The azides are explosive compounds. One has to keep the amounts of alkali azides in the decomposition tubes low in order to avoid hazards. The final step of this process is the vacuum distillation of the alkali metal product. The decomposition of azides proceeds according to equation (2).[2]

$$NaN_3 \rightarrow Na + \tfrac{3}{2}N_2 \quad (2)$$

The thermal reductive processes used to fabricate sodium metal are now only of historical interest. The same is true of the electrolysis of molten hydroxide (Castner process, 1891) and the electrolysis of rock salt using a binary salt mixture (Downs process, 1921). The modern production of sodium is based on the electrolysis of a ternary salt melt. In this process a modified version of the Downs cell is employed. The principle of this cell is similar to that of the electrochemical cell shown in Fig. 3.1. The molten electrolyte contains sodium chloride with small amounts of calcium and barium chloride in order to keep the calcium content of the final product below 1 wt%. The eutectic melt with 49 wt% $CaCl_2$, 31 wt% $BaCl_2$, and 20 wt% NaCl cannot be used because of its relatively high electrical resistance and the precipitation of large amounts of calcium.

The electrolysis is carried out in a continuous process in very large cells containing more than eight tons of the salt mixture. The voltage ranges from 6.3 to 7.0 V.

The anodes are made of graphite of high electrical conductivity, and the cell is made of the same materials as used in the case of lithium production. The product of electrolysis is somewhat impure, containing metals such as calcium and the oxides of calcium as well as sodium and some chlorides. The precipitation of all nonmetallic impurities and oxidation of the dissolved calcium metal is carried out by aging of the melt just above the freezing temperature at 110°C. All these precipitated compounds can be easily removed by filtering through steel mesh filters at the temperature of precipitation. The sodium produced and refined in this way still contains about 0.05 wt% calcium. Mild oxidation of the molten sodium at 300 to 400°C leads to further purification. Calcium is completely oxidized, and the oxide precipitates due to its very low solubility in the molten metal. The double-purified sodium is "reactor grade."

The rock salt required for the production of sodium metal has to be a very pure and dry raw material. It should be free from traces of heavy metals and magnesium. The crystal size of the salt is also important, since there should not be any salt dust on top of the molten mixture. Large crystals, however, often contain inclusions of rock salt solution and cannot be sufficiently dried. Rock salt alone has to be fed continuously into the cell to keep the bath composition constant. Chlorine is collected in a "dome" made of nickel, and the tubes needed for the transfer of the gas are fabricated out of steel.

Sodium is also formed as an intermediate in the Castner-Kellner process to produce chlorine and sodium by hydrolysis using amalgam cathodes. This process, however, is practically not applied for the production of sodium.

Sodium metal is available in the form of bars, wires, or hoops, or canned in tins of several sizes or in barrels. Even within the containers, protective gases (nitrogen or argon) or inert liquids (petrol ether) are required to prevent any reactions of the metal with the constituents of the atmosphere. Large quantities of sodium are filled into tanks which are mounted on trucks or wagons. They can have a capacity of up to 15 or 45 tons, respectively. Sodium is filled into these tanks in its molten state at about 120°C. The transport is undertaken after freezing the metal, since the solid form is less dangerous during transport. The tanks have provisions to be heated above the melting point of the alkali metal so that it can be pumped out. The most important producers and their annual production are listed in Table 3.1.

Table 3.1. Producers of Sodium Metal

Company	Country	Capacity (t/a)
Du Pont	USA	70,000
Ethyl Corp.	USA	68,000
Reactive Metals Inc.	USA	34,000
The Associated Octel Co., Ltd.	UK	20,000
Imperial Chemical Industries	UK	15,000
Métaux Spéciaux S.A.	France	12,000
Degussa	Germany	26,000
Nippon Soda Co., Ltd.	Japan	9000
		254,000

Though electrolytical processes are known for the reduction of potassium salts, in which potassium hydroxide or mixtures of chlorides, as in the case of sodium, are used, they have not been successful on an industrial scale on account of the fact that potassium has a higher solubility in the process melts than sodium, and that its higher vapor pressure causes reactions in the gas phase. The possible formation of potassium peroxide is also hazardous in the electrochemical process.

Potassium production of today is, therefore, based on thermal reduction processes, as for instance in the MSA (Mine Safety Appliances Corp., Callery, Pa., USA) process

$$KCl + Na \rightarrow K + NaCl \qquad (3)$$

in which potassium chloride is reduced by metallic sodium. The primary product of this chemical process is a sodium–potassium alloy. This alloy is the raw material for the production of a very pure potassium metal refined by vacuum distillation. Sodium–potassium alloys of desired composition are also produced by this method. The reaction has to be performed at a temperature high enough for the direct distillation of part of the potassium from the reaction mixture. A continuously working plant includes a rectification column which purifies the alkali metal to a sodium content of less than 1 wt%, sodium being the main impurity. A product of 99.99 wt% potassium is obtained by further purification.

Another thermal process (Griesheim) involves the reduction of potassium fluoride according to equation (4).

$$2KF + CaC_2 \rightarrow 2K + CaF_2 + 2C \qquad (4)$$

This reaction takes place in a brick furnace containing a steel chamber, in which the reaction mixture has to be heated to a temperature of 1000 to 1100°C. The potassium formed in the reaction is distilled into an iron tube, and the condensed liquid is protected by paraffin oil. Potassium fluoride is a rather expensive raw material; however, the process is simple and safe, and thus economical.

Other chemical reduction processes taking place at high temperatures and based on coal, iron, silicon, and other metals as reducing agents are not currently in industrial practice.

Following production of potassium by the processes described above, the soft and low-melting metal is cast into bars, which have to be carefully protected from contact with the air or humidity. Potassium–sodium alloys with 78 wt% potassium (melting point, 11°C; boiling point, about 785°C; density at 100°C, 0.847 g/cm^3) and with 56 wt% potassium (melting point, 19°C; boiling point, 825°C; density at 100°C, 0.886 g/cm^3) are also commercially available. They are stored and transported in containers of stainless steel with capacities of 1.5, 5, 12.5, or 100 kg. These metals cannot be stored in simple cans, since they are corrosive even at room temperature. Brazements made of tin alloys fail after short contact times due to the high solubility of tin in the potassium–sodium alloys.

The production of rubidium metal is coupled with the electrolysis of lithium, potassium, or cesium, since this element is always only a minor component in the ores of the other alkali metals. The main source for the production of rubidium is the residue of the carbonates from lithium production based on lepidolite. Rubidium has to be precipitated as Rb_2SnCl_6 or as $Rb_2Zn[Fe(CN)_6]$. The first complex is converted to RbCl by means of pyrolysis. The latter has to be thermally oxidized to form a more or less impure carbonate. The separation of rubidium from potassium or cesium is the main problem in these processes.

Separation by means of extraction with nitrobenzene, sulfophenols, or *n*-alkylphenols or with selective cation exchangers can help to solve these problems as can selective absorption by acid manganese dioxide serving as an ionic sieve.

The metal is produced by reduction of the salts in a vacuum device and final purification is achieved by vacuum distillation of the primary product. The electrolysis of molten salts is commercially unimportant.

Rubidium is the stable daughter element of the radioactive decay of ^{85}krypton, which is part of the radioactive waste in reprocessing plants for nuclear fuels. If the active gas is stored for a period of one hundred years, the inactive final product of the radioactive decay is the relatively precious alkali metal, which remains in the pressure cylinders. The metal contains only some oxide, hydroxide, or carbonate. Purification by means of vacuum distillation is the only necessary step for metal production.

The technical source for the production of cesium metal is pollucite, $2Cs_2O \cdot 2Al_2O_3 \cdot 9SiO_2 \cdot H_2O$. The mineral is treated with hydrofluoric acid in order to remove most of the silica. The residue of this treatment has to be dissolved in concentrated sulfuric acid. After the neutralization of the solutions, the heavy metals are precipitated as sulfides. Neutralization is performed through the addition of metallic aluminum. The product obtained by concentrating the solutions is cesium alum. The excess of aluminum is separated by precipitation, using barium hydroxide as the precipitating agent. Ammonium carbonate is used to remove the remaining barium hydroxide. The solution is finally concentrated in order to crystallize cesium carbonate. Other processes can be employed to obtain a pure salt as the raw material for the reduction. The choice of the process to be applied is strongly influenced by the initial composition of the ores.

One method to produce cesium is by the reduction of pure salts with reactive metals, for instance, calcium. The reaction is performed *in vacuo* at a temperature between 700 and 800°C. Under these conditions, cesium distills out of the mixture and is condensed as a relatively pure product. As in the case of other heavy alkali metals, cesium is sealed in glass ampoules, which are packed in adsorbing powders in order to prevent ignition in the event of breakage. Large quantities of the metal are cast in stainless steel containers under argon as cover gas.

For preparing rubidium in the laboratory scale, dry rubidium or cesium chloride together with pieces of calcium metal are placed in an iron cup, which is heated inside a glass apparatus. The reaction starts with the evolution of gases during the heating to 250°C. After the outgassing of the reaction mixture, the temperature is raised slowly to 600°C. The alkali metals distill off from the mixture and are refined by means of a second vacuum distillation step.[3] In larger-scale reactions the use of stainless steel vessels is recommended.

Purification by means of distillation *in vacuo* is an established laboratory method for the preparation of pure alkali metals. Pyrex glass apparatus is recommended for the distillation of sodium and the heavier alkali metals. The distillation of lithium requires vacuum distillation units made of metal, preferably iron or low-alloy steel.[4] In storage, the purified alkali metals must be protected from the atmosphere, and small amounts of the metals are stored in glass ampoules.

References

1. J. H. de Boer, J. Broos, and H. Emmens, *Z. Anorg. Allg. Chem. 191*, 113–121 (1930).
2. R. Suhrmann and K. Clusius, *Z. Anorg. Allg. Chem. 152*, 52–58 (1926).
3. L. Hackspill, *Helv. Chim. Acta 11*, 1008–1026 (1928).
4. G. Brauer, *Z. Anorg. Allg. Chem. 255*, 101–124 (1947).

4

Purification and Handling

4.1. Purification

The impurities normally found in alkali metals include both metals and nonmetals. The most abundant metallic impurities are other alkali metals, because they are often present in the ores and are fully miscible as liquid metals. Alkaline earth metals are somewhat soluble in alkali metals and may be present as impurities. Calcium is a particularly important impurity in sodium, arising from the electrolytic process itself. The coinage metals have some solubility in alkali metals and may be present as impurities if these metals have been used in any stage of the manufacturing processes. Boron and silicon are introduced as contaminants when liquid alkali metals come into contact with borosilicate glass. Most transition and heavy metals normally encountered in the handling of alkali metals have very low solubility and therefore are unlikely to be present in high levels. In general, commercially available alkali metals contain only very low levels (ppm) of metallic impurities. The exceptions are other alkali metals and alkaline earth metals. The former, if found undesirable for the particular application, have to be removed by special procedures such as vacuum distillation or oxidation. Calcium removal, being important for liquid sodium, is discussed separately below.

It is the nonmetallic impurities such as O, H, C, and N which are of greater concern in the purification and handling of alkali metals. Because of the reactive nature of these metals, they abstract these impurities from air, moisture, CO_2, oils, and greases. Oxygen would thus be present not only as dissolved oxide, but also as a surface coating (solid oxide). The case of hydrogen is similar. Even though carbon solubility is low, carbon can be present as dispersed particulates. Only lithium reacts strongly with nitrogen, forming a stable nitride which dominates the chemistry of this metal. Purification of alkali metals thus mainly involves removal of these nonmetallic impurities.

Purification requirements generally fall into three categories. First of all, the alkali metal procured from commercial sources must be purified before charging into various systems such as sodium loops. Depending upon the initial purity, as well as the desired final purity, purification schemes have to be devised. Secondly, continuous on-line purification is necessary for maintaining the purity of dynamic alkali metal systems. Thirdly, when small samples of highly pure alkali metals are required, special techniques have to be applied to attain and maintain the desired purity. Purification and handling of alkali metals has been discussed extensively by Mausteller *et al.*[1] Discussions on particular alkali metals or procedures are found in many specific references.[2-4]

The main techniques used in alkali metal purification are oxide slagging, filtration, cold trapping, and hot trapping. Both filtration and cold trapping take advantage of the low solubility of many elements, especially nonmetallic, in alkali metals just above the melting point. The impurities are filtered off in the initial charging stage in the former technique, while they are precipitated and retained in the cold leg of a loop in the latter case. The basis of the oxide slagging technique is that the oxides of many metals, such as calcium, are much stabler than the oxide of the alkali metal being purified. Reaction with controlled quantities of oxygen thus precipitates out such metal oxides which are then removed. In hot trapping, materials which have a greater affinity for the impurity than the alkali metal are used to remove the impurity. For example, zirconium is used as a getter for oxygen.

4.1.1. Filtration

Filtration removes both insoluble and precipitated impurities. It is the recommended procedure for purifying the as-received material and can be carried out during the process of transfer of the liquid metals to the system. Filtration is simple and can be performed at low temperatures. A temperature of about 15°C above the melting point is recommended for carrying out filtration.

A sintered stainless steel filter disk or a set of graded filters welded across a section of pipe is now commonly used for filtration. These disks are available in a variety of thicknesses and pore sizes and also in other metals such as nickel and monel. The lower the pore size, the greater is the pressure drop and the possibility of plugging. Thus low pore size (5 μ) filters are used only in the final purification step.

Filtration removes impurities which remain insoluble at the temperature of filtration. The oxygen levels in sodium can be brought down to a few ppm by this method and in potassium, to less than 10 ppm. Filtration is not considered suitable for the purification of lithium. Oxygen solubility in cesium and rubidium is so high that filtration is not a practical approach to their purification. Purification from hydrogen follows similar lines.

Carbon solubility is low at the filtration temperature, but carbon can exist as very fine particulates. Thus the degree to which carbon is removed can be quite unpredictable.

Many transition metals have very low solubility in alkali metals, while some other metals precipitate as oxides or nitrides. Thus several metals also get effectively removed in this purification step. In the purification procedures prescribed for sodium, initially the liquid metal is heated to 300°C to allow any Ca to react with Na_2O.

4.1.2. Cold Trapping

Cold trapping is a method of purification widely employed for the on-line purification of sodium in dynamic systems such as sodium loops. A cold point is established, usually in a bypass line, which causes the impurities to precipitate. The design of the cold trap ensures that a well-defined cold-point temperature is maintained and that precipitated impurities are retained. Cold trapping is an efficient technique by which oxygen levels of the order of 1 ppm can be achieved.

There are two types of cold traps—diffusion type and forced-circulation type. In the former a cold finger is established in a container that forms part of the liquid metal system. The impurities are allowed to diffuse to the cold point. The cold trap is located in the loop in such a way that natural circulation in the isothermal loop delivers the impurities to the mouth of the cold trap and thereafter the impurities move to the cold point by natural diffusion. In designing and operating such cold traps, care has to be taken to see that the mouth of the trap does not get plugged and that the trap is not overcooled to the point of solidification of the metal. A cold finger kept in a tank of molten sodium gathers a good deal of the precipitable impurities. This technique is often referred to as aging.

Forced-circulation cold traps are more commonly used in sodium loops. They are efficient devices which maintain quite accurately the purity levels in a dynamic sodium system when located on a bypass line. Figure 4.1 shows schematically a typical cold trap. It consists of a regenerative heat exchanger, a crystallizer tank, and a coolant system which maintains the trap temperature at a desired value. Stainless steel wire mesh is packed inside the crystallizer tank to aid crystallization by

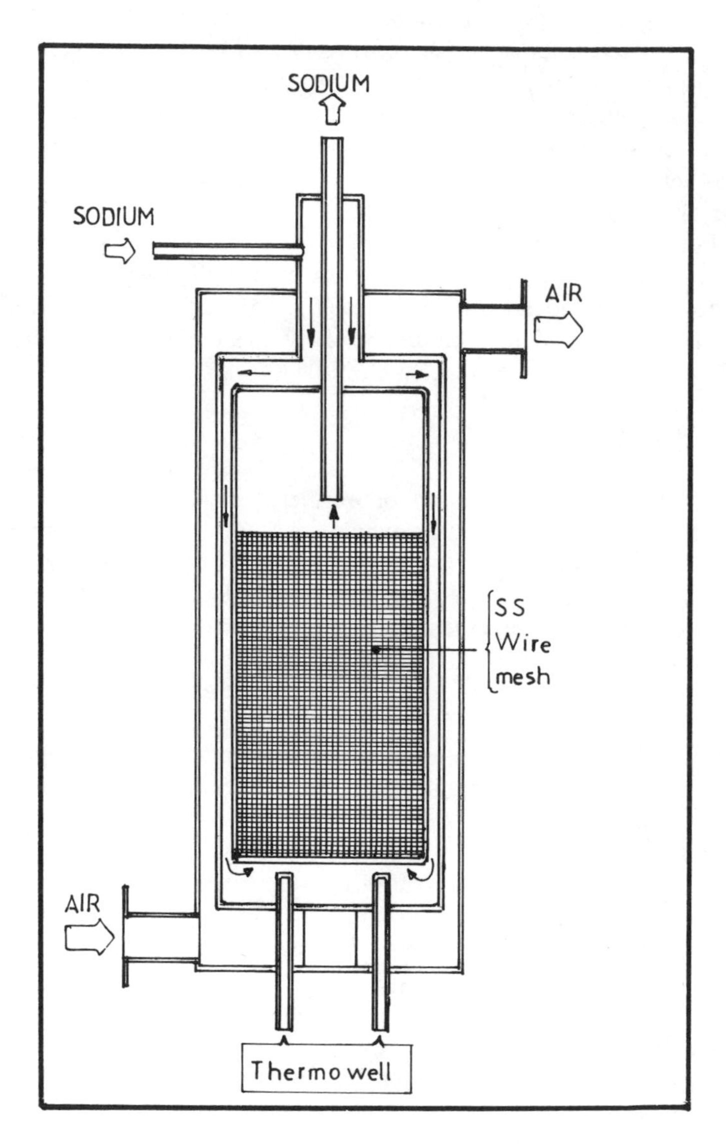

Figure 4.1. Schematic of a forced-circulation cold trap.

increasing the surface area and also to help retain the precipitated oxide in the trap. The coldest point in the trap determines the impurity level and must, therefore, be measured and used to control the cooling. Cooling fluids used include air, thermofluid, and NaK—the latter being particularly efficient. The most commonly used cooling system for cold traps is a NaK jacket with embedded cooling coils through which a thermofluid is circulated to remove heat.

The time required for the system to achieve the desired level of purity depends on the nature and level of the impurities and the fraction of the sodium which passes through the trap in unit time. If C_0, C_t, and C are the oxygen contents at times 0, t, and ∞ (i.e., after equilibrium is achieved), they are related to the cold-trap flow rate w (kg/h) and total system inventory W (kg) by the following expression:

$$\frac{C - C_\infty}{C_0 - C_\infty} = e^{-(w/W)t} \qquad (1)$$

There are many sodium loops operating, including in fast reactors, in which oxygen levels are maintained at less than 5 ppm by means of cold traps.

Cold traps remove all precipitable impurities. They are particularly efficient in the removal of hydrogen because of the very low solubility of hydrogen at cold-trap temperatures and the favorable kinetics of hydride precipitation. They also remove carbon, but the efficiency depends on the particle size of the dispersed carbon. Insoluble corrosion products also get collected in cold traps.

Cold trapping has been used in the purification of potassium to oxygen levels of less than 100 ppm. It is ineffective for the purification of rubidium and cesium.

4.1.3. Hot Trapping

Impurities such as oxygen, hydrogen, carbon, and nitrogen can be removed from alkali metals by converting them to insoluble compounds with metals which have high chemical affinity for them. The reaction (gettering) is carried out at fairly high temperatures and hence this technique is known as hot trapping.

The getter metal that is used may be either soluble or insoluble in alkali metals. Calcium and magnesium are examples of soluble getters. Yttrium has been used as a getter in lithium loops. The oxide (or other compounds) that is formed must be removed, for example, by filtration.

Zirconium and uranium have been used for gettering oxygen in sodium. In a typical hot trap, sodium flows through a vessel packed with zirconium chips kept at 600–650°C. Sodium oxide is reduced to sodium by zirconium with the formation of ZrO_2 which remains adherent. Oxygen levels of less than 1 ppm can be obtained by this method.

4.1.4. Distillation

Distillation can purify alkali metals from a large number of impurities, especially metals. This method is more commonly used in the purification of potassium than that of other alkali metals. Potassium can be vacuum distilled in the temperature range of 250–300°C. Distillation of lithium requires temperatures in excess of 500°C while sodium is distilled in the temperature range of 350–400°C. Rubidium and cesium can easily be vacuum distilled to remove many metallic impurities.

Metals which have appreciable vapor pressure at the temperature of distillation are not easily removed by this process (e.g., Zn, Cd). Distillation is effective in removing oxygen, but special care must be taken to avoid contamination by the cover gas of the distillate and the adsorbed moisture on the walls of the receptacle.

Distillation of sodium is an energy-intensive process requiring considerable capital expenditure to achieve large throughputs.

4.1.5. Oxide Slagging

Metals having a greater affinity for oxygen than does the alkali metal of interest can be removed by adding controlled quantities of oxygen to the alkali metal and heating. The precipitated oxides can be removed by filtration. Sodium and potassium are conveniently purified in this manner from impurities such as calcium, barium, magnesium, lithium, and several transition metals. The method has very limited application to lithium because of the high stability of Li_2O.

4.1.6. Recommended Purification Methods

The purification method that is chosen depends on the alkali metal, the nature and level of the impurities, and the purity required. Alkali metals are now commercially available at acceptable purity levels. Purification procedures must mainly remove impurities like oxygen, hydrogen, and carbon, which may be picked up on handling, and alkaline earth metals, which may not be fully removed in the commercial purification steps.

There is a great abundance of experience available on sodium purification. Filtration followed by cold trapping is the common procedure adopted. Oxide slagging can be used as a first step if calcium content is high. Potassium can also be purified in a similar manner, though distillation is also employed. For small samples the authors have found that oxidative slagging followed by filtration in a high-purity glove box is adequate. To achieve very high purity, especially in oxygen, hot trapping may be used. Distillation is particularly effective in removing carbon.

Nitrogen is the principal impurity of concern in lithium systems. Hot trapping with titanium at about 700°C is effective in removing nitrogen. Filtration followed by hot trapping with zirconium is recommended for the purification of rubidium and cesium.

4.2. Handling

As alkali metals are highly reactive, care must be taken in handling them. Handling techniques must ensure the safety of the operator, on the one hand, and the maintenance of adequate purity, on the other. Handling of liquid metals requires greater care and sophistication than handling of the solid form.

Small quantities of sodium are usually obtained in the form of bricks. These bricks are completely coated with white sodium oxide and are transported in gas-tight drums. Very small quantities for laboratory use can be obtained in sealed cans.

Potassium is generally received as small bricks or cylinders coated with paraffin. The paraffin may be removed by washing with *n*-hexane. Potassium must be handled with greater care than sodium, as the superoxide coating can lead to explosions. Rubidium and cesium are normally received in sealed bottles. They are best handled in inert-atmosphere glove boxes.

Lithium is normally available in the rod form in sealed containers. These rods should preferably be handled in an argon atmosphere for sampling as an adherent nitride coating is formed on exposure to air. However, lithium can be cut in air also as it does not catch fire on exposure.

Sodium bricks can be cut in air using an oil-coated knife or a guillotine-type slicer. The oil film left behind on the fresh surfaces will limit the oxygen pickup. The precautions to be taken while cutting sodium in air are:

1. Thin latex gloves should be worn over cotton gloves to protect the hands. Sodium metal should not be touched with bare hands.

2. Suitable protective shoes, face masks, and laboratory apparel should be worn.
3. Sodium should preferably be cut in a laboratory with low humidity.
4. Eyewash stations should be accessible for use in an emergency.
5. The waste sodium chips should be promptly transferred to a waste sodium container covered with a lid.

Potassium metal catches fire on exposure to air. So alkali metals like potassium, rubidium, and cesium should necessarily be handled in inert-atmosphere boxes. Even inside the box, they should not be left exposed for a long period as a superoxide coating will be formed. They should be stored in sealed vessels.

All alkali metals are more reactive in the liquid than in the solid state. For heat transfer applications the liquid alkali metal must be in a state of high purity. Liquid metal systems such as sodium loops and storage tanks are ultra-clean highly leak-tight systems with a high-purity inert cover gas maintained over the free liquid surface. When analytical, chemical, and metallurgical investigations are carried out, the sodium samples must have representative purity. This demands handling facilities such as high-purity inert-atmosphere glove boxes and sampling, transfer, and storage techniques.

4.2.1. Inert-Atmosphere Glove Boxes

When small quantities of alkali metals are handled for various investigations, the atmosphere has to have very high purity (<1 ppm in O_2 and H_2O). This is accomplished in an inert-atmosphere glove box fitted with a recirculatory purification system. The gas to liquid metal ratio in a glove box is much higher than in a sodium loop or a large storage tank and hence the requirement for purity is higher.

The inert gas normally used nowadays is argon. Several purification schemes have been used for the recirculatory purification of the cover gas. A NaK bubbler used to be a popular choice. However, its use has been largely discontinued for several reasons such as aerosol formation and plugging of lines. In addition, NaK has to be frequently replaced as it gets consumed. Removal of moisture by the use of molecular sieves followed by hot gettering with zirconium, titanium, or uranium has been adopted by some groups. At present, most workers prefer the use of molecular sieves (Linde 4A, 13×) for moisture removal and a copper-based catalyst (e.g., BASF R3-11) for oxygen removal as such a purification unit can be readily regenerated and operates at room temperature. For handling lithium it is essential to remove even traces of nitrogen from the atmosphere. This can

Figure 4.2. Commercial dry argon glove box. (Courtesy Mr. Hinrichs, M. Braun, Garching, Germany.)

be achieved by introducing a nitrogen trap in the recirculation line in the form of a titanium sponge heated to about 650°C.

Inert-atmosphere glove boxes for liquid metal handling are commercially available from M/s. Vacuum Atmospheres, USA, and M. Braun, West Germany (Fig. 4.2). Figures 4.3 and 4.4 show the schematic and a photograph of the sodium handling box in use in the laboratories of one of the authors. It consists of a cylindrical stainless steel box designed to be very highly leak-tight. The glove ports are fitted with butyl rubber gauntlets. A diaphragm pump continuously circulates the argon gas through a purification tower filled with pellets of BASF R3-11 catalyst and

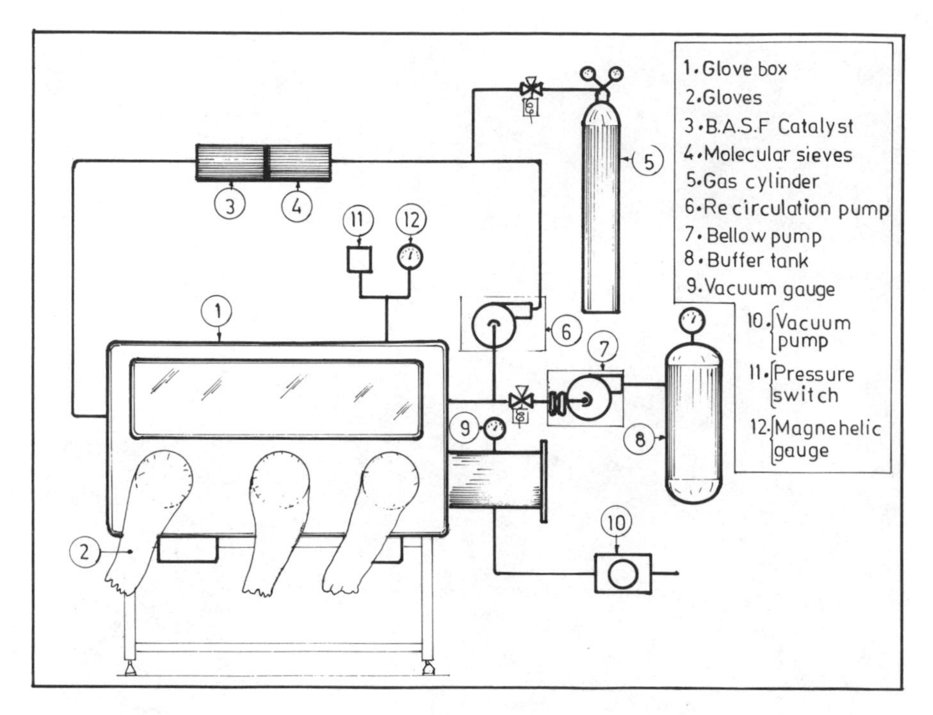

Figure 4.3. Schematic of an inert-atmosphere box for sodium handling.

Linde molecular sieves 4A. A sodium bath kept at about 120°C inside the box serves as a good indicator of the purity level. If liquid sodium retains a scum-free surface for a few hours it means that oxygen and moisture are present at sub-ppm levels. The main impurity ingress is moisture entering through the gloves. It is, therefore, necessary that hands be covered by cotton gloves, and sweating of hands is to be avoided during the operation of glove boxes. As a consequence of moisture ingress, hydrogen partial pressure inside the box may be high unless specifically controlled.

4.2.2. Sodium Loops

When liquid metals are used under dynamic conditions as heat transfer fluids or in the study of corrosion and material behavior, they circulate in liquid metal loops. In a fast breeder reactor, the mass flow is large ($\sim 10{,}000\ m^3/h$), and the flow velocity is about 7 m/s. The experimental loops have much lower flow rates (a few cubic meters/h) but simulate the flow velocity and temperature conditions.

Figure 4.4. Photograph of an inert-atmosphere box used for sodium handling.

A typical experimental sodium loop is designed as a "figure-of-eight" loop. Such a loop is schematically shown in Fig. 4.5. The test section is located on the hot leg, whereas the pump, the cold trap, and the flow meter are located on the cold leg. The two legs meet at the economizer. The experimental loops use electromagnetic pumps while in the large reactor loops centrifugal pumps are used. A more detailed description of sodium loops and their components, such as pumps and flow meters, is beyond the scope of this book. Argon is generally used as cover gas.

4.2.3. Wetting by Liquid Alkali Metals

When a liquid is used as a heat transfer fluid, its intimate contact with the pipe carrying it is essential for efficient transfer of heat across the solid–liquid boundary. Similarly, in electromagnetic pumps and in magnetohydrodynamic devices one must ensure good electrical conductivity across the interface between the solid and the liquid phases. The presence of gas bubbles and their spread in liquids having poor wetting properties are quite undesirable. For example, consider an electromagnetic pump for sodium. Here, a high electric current is passed across a sodium-

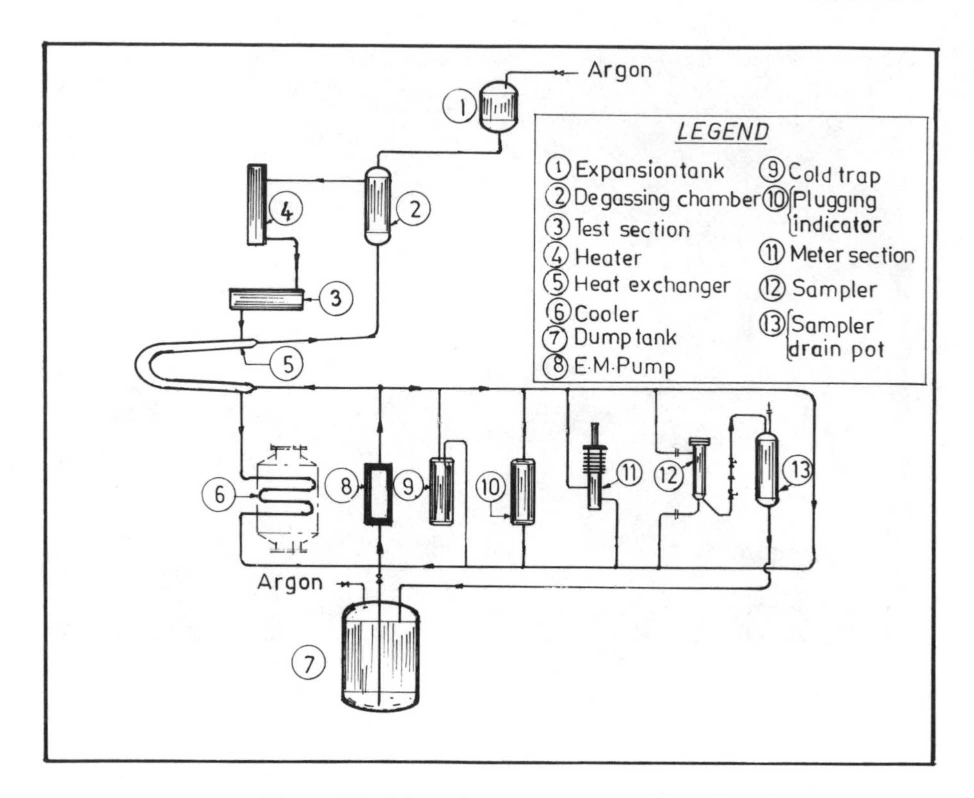

Figure 4.5. Schematic diagram of a sodium loop.

containing duct which is in a magnetic field whose direction is perpendicular to both the current and the sodium flow. In the absence of good wetting, electrical resistance at the sodium–duct interface would be large, thus significantly reducing pumping action. If the current is high, it bypasses the sodium and flows through the duct wall, causing it to be burned out.

Wetting of metals by liquid sodium is related to the presence of surface films on the metals and their reactions with the liquid metal. Pure sodium readily spreads over the surface of pure metals if surface films are absent. As transition metals carry an invisible film of oxide even when polished in air, it is the reaction between the surface film and sodium that determines the solid metal's wettability. If the oxide is less stable than sodium monoxide, as is the case with Fe, Cr, and Ni, it is reduced to metal by sodium with the result that the clean metal surface is readily wetted. When the free energies of formation of the oxides of the solid and liquid metals are similar, then there is the possibility of ternary oxide formation, which also promotes wetting. When the oxide film is stable in sodium, then it is the wettability of the oxide itself that must be considered. The behavior of liquid potassium must be very similar to that of sodium in view of the similarity of its reactions with transition metal oxides to those of sodium. Lithium oxide being much more stable than other alkali metal oxides, lithium can reduce most metal oxides to the metallic state. Thus, liquid lithium would wet transition metal surfaces more readily than other liquid alkali metals.

Wetting studies carried out by measuring the contact angle between liquid and solid phases have established a "critical wetting temperature" characteristic of each metal, above which complete wetting takes place.[5] For wetting by liquid sodium, iron has a critical wetting temperature of 140°C while cobalt and nickel exhibit the same phenomenon at about 190°C. The critical wetting temperatures of molybdenum and tungsten are about 160°C while that of chromium is somewhat higher. This behavior can be explained on the basis that the metal oxide film is reduced by sodium above the critical wetting temperature. In the case of chromium, it must be the formation of $NaCrO_2$ which causes wetting. In the case of refractory metals such as titanium, zirconium, and vanadium, there is no critical wetting temperature. Wetting, however, takes place at sufficiently high temperatures (>200°C). Stainless steel, which is the material generally used to contain liquid sodium, also gets wetted at similar temperatures.

4.2.4. Sampling

Collection of a representative sample of a liquid alkali metal requires special techniques. The process of sampling and the subsequent handling and transport of the sample must not introduce impurities. Extreme care has to be taken to keep out atmospheric contaminants such as oxygen, nitrogen (in the case of lithium), carbon dioxide, and moisture. Sampling vessels, unless carefully chosen and scrupulously cleaned, especially from moisture and carbonaceous materials, can also contribute impurities. The following precautions must be taken to obtain representative samples:

1. Sampling should be done at the operating temperature of the alkali metal system, which is normally high (e.g., 400–500°C in sodium systems). Since most of the impurities have highly temperature-dependent solubility, even a small reduction in temperature may alter the impurity levels.

2. Most of the impurities segregate strongly during cooling of the liquid metal sample. This makes it necessary to analyze the entire sample for a given impurity (e.g., O, H, and C in sodium).
3. Suitable crucible material should be used while sampling for various impurities as the crucible can collect the impurity and influence the analytical results (e.g., Ni cannot be used in sampling for trace metals and radionuclides as it is a good trap for many of these metals). The desirable materials for various impurities in sodium are listed below:

   Oxygen and hydrogen: Nickel crucible
   Carbon: Alumina crucible
   Trace metals: Tantalum, titanium, or molybdenum
   Radionuclides: Stainless steel/quartz
   Tritium: Quartz

4. The blank introduced by the crucibles has to be taken into account when sampling for nonmetals. Proper cleaning of the crucible and adequate flushing with hot sodium while sampling have to be carried out.
5. Transport containers and handling devices should be so designed as to minimize impurity ingress. It is desirable to minimize handling steps.

One of the early methods of sampling was dip sampling. Here, a sample bucket is lowered into the liquid metal to withdraw a sample. A typical sampling station would consist of an inert-atmosphere box from which the bucket can be lowered into the liquid metal through a ball valve. The sample is withdrawn into the box and then appropriately handled. It is here that contamination can take place unless extreme care is taken. If the sample has to be transported to the laboratory or temporarily stored, these steps are likely to introduce contaminants, especially oxygen and hydrogen.

Another method of sampling used in sodium loops may be termed bypass tube sampling. Here the liquid metal is allowed to flow through a thin-walled nickel tube in a bypass line. After sufficient flushing is established at an appropriate temperature, the nickel tube is cooled and crimped near the ends. It is then cut outside the crimps and these tube samples are supplied to the laboratory for analysis. The analyst can cut out a portion of this tube and extrude the alkali metal under inert atmosphere for analysis.

Tube samples can be conveniently taken but may lead to erroneous results for impurities which segregate on cooling. As the cooling is from outside, several impurities tend to concentrate at the periphery and may stick to the tube walls. Thus the extruded sample is not representative.

4.2.4.1. Overflow Sampling

If the liquid metal is allowed to fill and overflow a sample cup placed in a bypass line and the whole sample thus collected is thereafter used in the analysis, then the problem associated with tube samples can be avoided. The "Harp sampler" developed at TNO* is the best example.[6]

In this method, (i) it is possible to flush the crucibles; (ii) different crucibles can be used for different impurities; (iii) sampling can be done at the system temperature; (iv) samples suitable for analyzing the entire sample for any impurity are obtained; and (v) there is a built-in sealed container for transporting the sample to the laboratory.

Figure 4.6 schematically represents the harp sampler. The clean harp is connected to the sampling line using Conoseal coupling. The harp is evacuated. The suction valve is then closed and the sampler is heated to the system temperature. The sodium valves are opened and flow is effected. After flushing, the sodium valves are closed and the excess sodium is sucked out into a chamber by opening the suction valve. The harp is cooled by opening the insulation. The harp is removed after the Conoseal area has been cooled and the seals disconnected. The sampler now becomes a transport container with the sodium freeze seals preventing any ingress of air or moisture. It can be opened in an inert-atmosphere box and each sample fully used for a given analysis. The TNO harp is probably the most widely used sampling device at present.

4.2.5. In-Line Distillation Sampling

A convenient method for sampling which minimizes impurity pickup is in-line distillation sampling. In this method, an alkali metal sample of known amount is collected and distilled in-line and only the residue is taken out for analysis. This may avoid the delays involved while collecting samples from radioactive systems and also the problems associated with waste dispersal. The problems of atmospheric contamination are solved altogether; larger samples can be taken, thus enhancing the reliability of the analytical results.

* Central Technical Institute, TNO, Apeldoorn, Netherlands.

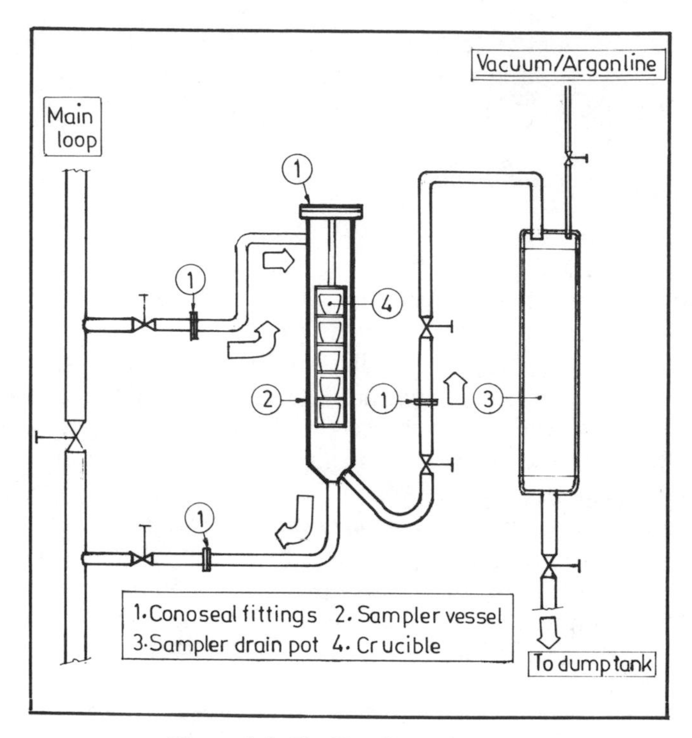

Figure 4.6. The "harp" sampler.

Such a method has been developed at the Argonne National Laboratory for use in EBR-II.[7] Figure 4.7 gives a diagram of the assembly used. A molybdenum cup of large volume ($\sim$50 ml) is connected in the line using a high-temperature metal seal. The crucible zone is evacuated and the lines are heated to effect sodium flow. The excess sodium is pushed back by applying pressure. This is then evacuated to distill the sodium. The distillate collects in the return line, which is not heated. After the distillation is over, the crucible is removed and the residue is analyzed. The harp sampler has also been modified for in-line distillation.[8]

4.3. Sodium Fires

Alkali metals react with oxygen and moisture in air. This reaction can lead to alkali metal fires. Sodium fires have been investigated in detail on account of the potential for their occurrence in sodium-cooled fast breeder reactors.

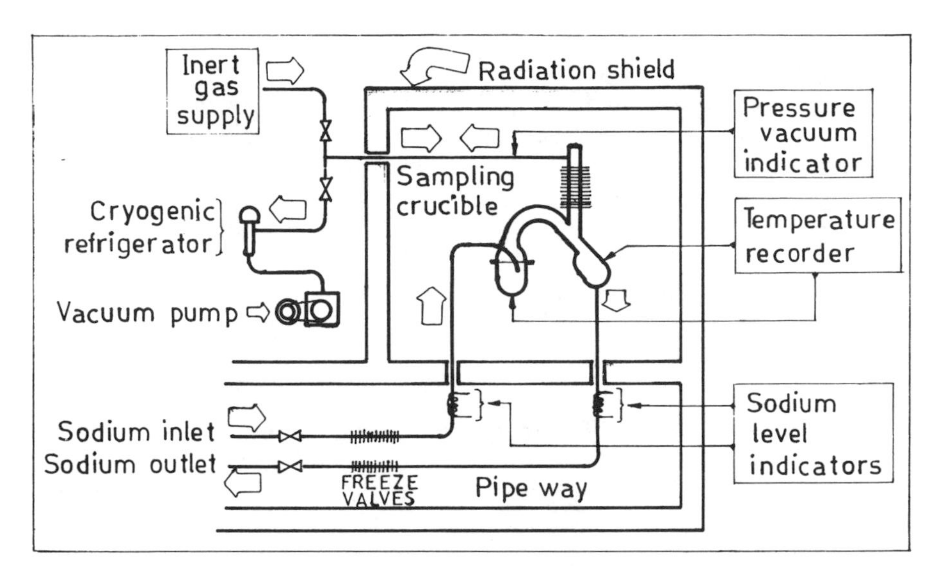

Figure 4.7. Schematic of an in-line distillation setup.

Solid sodium does not react with oxygen or moisture vigorously. The solid products of the reaction, Na_2O and NaOH, form a protective coating over the solid bricks, preventing further reaction. Therefore, sodium bricks are safely handled in the laboratory.

Liquid sodium reacts more vigorously with oxygen and moisture, but ignition is not attained until a self-sustaining temperature rise is established. In other words, ignition temperature is that temperature at which the oxidation becomes sufficiently rapid that a bulk temperature rise occurs. Ignition temperature depends on several factors such as surface-to-volume ratio and oxygen and humidity content in the atmosphere. Thus the ignition temperature for a sodium spray is much lower than that for a pool of sodium. The former is generally around 120°C for sprays from sodium loops but could even be lower under fine mist conditions.

The ignition temperature of a sodium pool is 200–320°C depending upon the moisture content of the atmosphere. If the pool is stirred, ignition takes place at lower temperatures. Once ignition has started, combustion proceeds through a surface oxidation process which creates a solid layer on the pool surface. This surface layer thickens rapidly with oxide nodules or pillars growing in random positions. As the bulk sodium temperature rises to 400–500°C, vapor-phase combustion begins on the oxide pillars which act as wicks for the liquid metal. When the bulk temperature further increases to 600–650°C, the oxide wicks disappear in the pools and at still higher temperatures burning occurs from the metal surface.[9]

The aerosol released from pool and spray fires in oxygen-containing atmospheres ($>10\%$ O_2) consists mainly of Na_2O_2. This can be further transformed to NaOH and Na_2CO_3 depending on the availability of H_2O and CO_2. The reaction product in a sodium pool is mainly the monoxide (which is the oxide stable in equilibrium with the metal) and smaller quantities of NaOH, Na_2CO_3, and Na_2O_2.

Ignition temperatures of potassium and NaK are lower than that of sodium. Higher alkali metals should ignite at still lower temperatures. Lithium burns in nitrogen also, if the temperature is high enough, with the formation of Li_3N.

4.3.1. Prevention and Fighting of Sodium Fires

Unlike oil fires, metal fires do not require sparking and hence cannot be blown out. Liquid sodium and oxygen react spontaneously above the ignition temperature. The flame cannot easily be chemically quenched.

In order to prevent sodium fires, the liquid metal must be handled in an inert environment. Leaks from liquid sodium containers, especially sodium loops, cannot be ruled out and, therefore, precautions should be taken to prevent a fire in the event of a leak. The containment system around a sodium vessel or pipe work may be permanently under an inert atmosphere or provision may be made for inert gas flooding at the time of a fire.

A catch tray is sometimes used to collect the sodium that leaks out. The tray, which has sufficient volume to collect the leaked sodium but not much excess volume, is covered with baffles with sloping sides to facilitate the draining of sodium. The opening available for this purpose is very small compared to the surface area of the tray. In such a catch tray a sodium fire gets extinguished as the fire becomes starved of oxygen.

One approach to extinguishing a sodium fire is to exclude oxygen from the burning surface by covering it with a suitable powder.[9,10] Sand, sodium chloride, sodium carbonate, and sodium bicarbonate have been succesfully used to put out small fires. The powder must be free-flowing. It is a good practice to keep such powders handy to put out small sodium fires. The disadvantages of these materials are the large ratio of powder to the sodium in the pool that is required and the tendency of the powder to sink in the sodium. Recently powders based on graphite have been found to be far superior because they float on sodium and are effective in much smaller quantities. Graphex is an example and a 0.6-cm layer of Graphex (unexpanded) has been reported to extinguish sodium fires at 650°C.

Lithium fires are more difficult to extinguish than sodium fires. Since lithium burns in nitrogen, oxygen-starvation methods do not work with lithium. Further, lithium reacts with nearly all powder extinguishers. Liquid lithium being lighter than the powders, it would continue to rise above the powder, thus making this method of extinguishing also ineffective. Graphite-based powders may be used for extinguishing a lithium fire but repeated application of the powder is necessary. Making the atmosphere inert with argon is also effective if the fire is in a confined volume.

References

1. J. W. Mausteller, F. Tapper, and S. J. Rodgers, *Alkali Metal Handling and Systems Operating Techniques*, Gordon and Breach, New York, 1967.
2. A. W. Thorley and A. C. Raine, in: *Alkali Metals*, Spec. Publ. No. 22, The Chemical Society, London, 1967, pp. 374–392.
3. D. F. Shriver, *The Manipulation of Air-Sensitive Compounds*, McGraw-Hill Book Company, New York, 1969.
4. M. Sittig, *Sodium—Its Manufacture, Properties and Uses*, Reinhold Publishing Corporation, New York, 1956.
5. C. C. Addison, *The Chemistry of the Liquid Alkali Metals*, John Wiley and Sons, Chichester, 1984.
6. J. F. M. Rohde, M. Hissink, and L. Bos, *J. Nucl. Energy 24*, 503–508 (1970).
7. W. H. Olson, *Nucl. Technol. 12*, 7–11 (1971).
8. L. Bos, *J. Nucl. Energy 24*, 607–608 (1971).
9. R. N. Newman, *Progress in Nuclear Energy 12*, 119–147 (1983).
10. C. Raju and R. D. Kale, Report RRC-34, Reactor Research Centre, Kalpakkam, India, 1979.

5

Oxygen in Alkali Metals

5.1. Oxides of Alkali Metals

Alkali metals form a range of oxides: monoxide, peroxide, superoxide, and ozonide. When the metals react with oxygen, the prevailing conditions of temperature and oxygen potential determine which oxide is formed. It is the monoxide, M_2O ($M =$ alkali metal), that is generally present in equilibrium with liquid metals. Lower oxides than M_2O are known in the case of rubidium and cesium, but they have low (<445 K) melting temperatures and are not very important from the point of view of technology. The monoxides, however, are frequently encountered in liquid metal systems. When sodium burns in dry air or oxygen, sodium peroxide (Na_2O_2) is produced. Lithium also forms the peroxide with dry oxygen, but when it burns in dry air lithium nitride forms in greater abundance. When potassium, rubidium, or cesium burns in dry air or oxygen it is the superoxide, MO_2, that is produced. Thus in the case of the heavier alkali metals, the superoxide must be considered as the thermodynamically most stable oxide. For sodium the peroxide is more stable, while lithium is not known to form the superoxide.

5.1.1. Monoxides

Sodium monoxide can be prepared by the action of a limited amount of dry air on sodium

$$4Na + O_2 \xrightarrow[\text{(dry)}]{\Delta} 2Na_2O \tag{1}$$

Table 5.1. Properties of Alkali Metal Oxides[a]

Alkali metal oxide	Li	Na	K	Rb	Cs
M_2O					
M.p.	1843[b,c]	1405[b]	1154[c]	673[b,c]	763[c]
Density	2.013[c]	2.27[c,d]	2.32[c,d]	3.72[c,d]	4.25[c]
Crystal form	Cubic	Cubic	Cubic	Cubic	Rhombohedral
$\Delta_f G^0_{298}$	−562.1	−378.3	−320.3	−302.4	−308.4
M_2O_2					
M.p.	713[c]	948[b]	763[b,c]	843[b,c,d]	863[d]
Density		2.80[c]	2.40[d]	3.65[c]	4.74[d]
Crystal form		Hexagonal	Orthorhombic	Orthorhombic	Orthorhombic
$\Delta_f G^0_{298}$	−571.0	−448.2	−429.8	−405.2	−417.7
MO_2					
M.p.	1843[b]	825[b]	653[c,d]	685[b,c]	705[b,d]
Density		2.21[d]	2.14[c,d]	3.06[d]	3.80[d]
Crystal form	Cubic	Orthorhombic	Tetragonal	Tetragonal	Tetragonal
$\Delta_f G^0_{298}$	−562.1[b]	−222.5	−240.6	−244.9	−252.0

[a] Melting point in Kelvin, density in g/cm³, and Gibbs' free energy in kJ/mol.
[b] Data from ref. 3.
[c] Data from ref. 4.
[d] Data from ref. 2.

In fact, the solid oxide that separates from liquid sodium maintained under inert or low-oxygen environment is the monoxide. The monoxides of rubidium and cesium cannot be prepared by the partial oxidation of the metal, because the oxidation proceeds directly to the superoxide. In the case of potassium also, oxidation readily leads to the formation of the peroxide. However, partial oxidation followed by equilibration with excess potassium ensures the absence of the peroxide. The monoxide is obtained by distilling off excess potassium. All these monoxides can, however, be obtained by the reaction of the metal with the nitrate

$$10M + 2MNO_3 \rightarrow 6M_2O + N_2 \qquad (M = Na, K, Rb, Cs) \qquad (2)$$

Sodium and lithium peroxides, when heated in an inert atmosphere, yield the monoxides. Rubidium monoxide disproportionates into rubidium peroxide and rubidium metal at 683 K.[(6)] Some physical properties of the monoxides are given in Table 5.1.

5.1.2. Peroxides

Sodium peroxide is readily prepared by burning sodium in dry air. Lithium peroxide is produced by reacting lithium hydroxide monohydrate with 30% hydrogen peroxide; the resulting lithium hydroperoxide is filtered and vacuum dried to yield the peroxide. Potassium, rubidium, and cesium peroxides can be prepared by the controlled oxidation of the metal, but this does not yield the pure product, as some superoxides are also formed. These peroxides can also be obtained by the careful oxidation of a liquid ammonia solution of the metal with a stoichiometric quantity of oxygen at low temperatures. For example, K_2O_2 is prepared by the oxidation of a liquid ammonia solution of potassium at -50 to $-60°C$. The thermal decomposition of the superoxide also yields the peroxide; e.g.,

$$2KO_2 \xrightarrow{400°C} K_2O_2 + O_2 \qquad (3)$$

Some properties of the alkali metal peroxides are listed in Table 5.1.

*5.1.3. Superoxides**

The superoxides of potassium, rubidium, and cesium are readily obtained by the oxidation of the metals in dry air or oxygen. To prepare sodium superoxide, the peroxide must be heated in oxygen under pressure,

$$Na_2O_2 + O_2\ (150\ \text{atm}) \xrightarrow[100\ h]{450°C} 2NaO_2 \qquad (4)$$

Lithium superoxide is not known.

The superoxides decompose on heating to give oxygen. Even reactions with moisture and acids release oxygen:

$$2MO_2 \xrightarrow{\text{heat}} M_2O_2 + O_2 \qquad (5)$$

$$4MO_2 + 2H_2O \longrightarrow 4MOH + 3O_2 \qquad (6)$$

The main application of the superoxides is in the oxygen mask, where cannisters of KO_2 were originally used. It reacts with the exhaled breath of the wearer to generate oxygen.

$$4KO_2 + 2CO_2 \rightarrow 2K_2CO_3 + 3O_2 \qquad (7)$$

$$4KO_2 + 2H_2O + 4CO_2 \rightarrow 4KHCO_3 + 3O_2 \qquad (8)$$

More recently sodium peroxide is used in the masks as it is cheaper.

Some properties of alkali metal superoxides are listed in Table 5.1.

* According to IUPAC resolution G-9, these oxides should be called hyperoxides, but the more familiar name is retained here.

5.1.4. Ozonides

The ozonides of alkali metals are formed when the dry hydroxide is exposed to ozone:

$$3MOH(s) + 2O_3(g) \rightarrow 2MO_3(s) + MOH \cdot H_2O(s) + \tfrac{1}{2}O_2(g) \qquad (9)$$

These are bright-colored compounds. They undergo hydrolysis, liberating oxygen:

$$4MO_3 + 2H_2O \rightarrow 4MOH + 5O_2 \qquad (10)$$

5.1.5. Other Oxides

Some other oxides of alkali metals are reported in the literature. An example is the sesquioxide, M_2O_3, reported for potassium, rubidium, and cesium. But they are probably only combinations of other oxides.

Oxides lower than the monoxide have been reported to exist in the Rb–O and Cs–O systems. The phase diagrams of the Rb–O and Cs–O systems given in Figs. 5.1a and 5.1b are based on Refs. 6 and 5, respectively. Rubidium forms Rb_3O which incongruently melts at 321 K according to the following equation:

$$Rb_3O \rightarrow Rb_2O + Rb \qquad (11)$$

Four oxides are indicated between Cs and Cs_2O: Cs_7O, Cs_4O, Cs_2O_7, and Cs_3O. All of them melt at low temperatures and their properties are not well established.

5.2. Oxygen Solubility

Oxygen dissolves in liquid alkali metals, its solubility being a strong function of temperature. Oxygen solubility varies widely among the alkali metals; it is lowest in the case of lithium and sodium. Oxygen solubility in higher alkali metals increases with increasing atomic number. When the solubility is exceeded the monoxide (M_2O) generally precipitates. (As discussed earlier, the lower oxides may precipitate at relatively low temperatures in the case of Cs and Rb.) The equilibrium between the solid monoxide and the liquid metal at a given temperature is characterized by a unique oxygen potential ($\overline{\Delta G_{O_2}}$) which is related to the free energy of formation of the monoxide, $\Delta_f G_T^0$.

$$2M(s) + \tfrac{1}{2}O_2(g) \rightleftharpoons M_2O(s) \qquad (12)$$

$$\overline{\Delta G}_{O_2} = RT \ln p_{O_2} = 2\Delta_f G_T^0 \qquad (13)$$

(a)

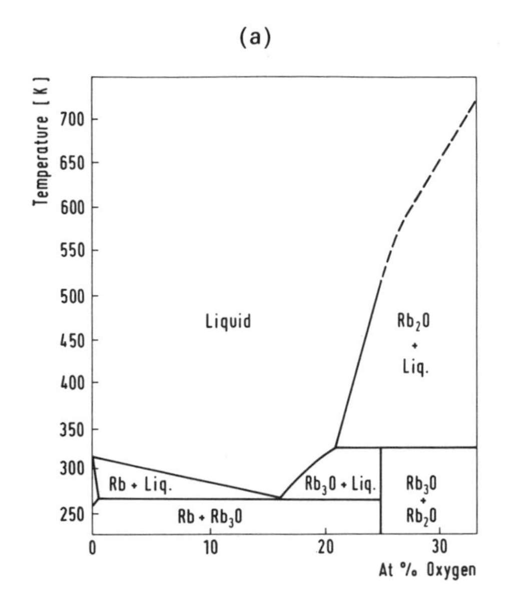

(b)

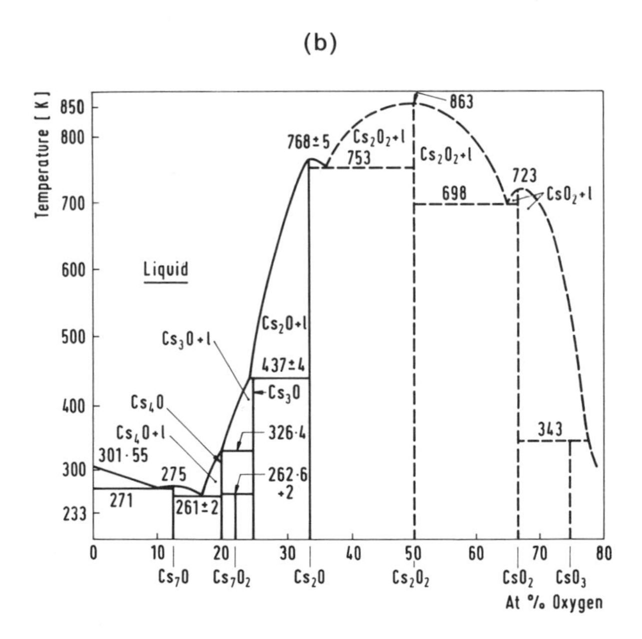

Figure 5.1. (a) Rubidium–oxygen phase diagram (after ref. 6). (b) Cesium–oxygen phase diagram (after ref. 5).

Considering the monoxide phase as the standard state for oxygen, the saturated solution which is in equilibrium with it must also have unit activity with respect to oxygen. Assuming Henry's law to be valid, the activity of the solution below the saturation level is given by the ratio of the oxygen concentration (C_O) to the saturation solubility ($C_{O,sat}$).

$$a_O = C_O / C_{O,sat} \tag{14}$$

The oxygen potential of such a solution is given by

$$\overline{\Delta G}_{O_2} = 2\Delta_f G_T^0 + 2(\ln C_O - \ln C_{O,sat}) \tag{15}$$

Thus, in order to compute the oxygen potential of a liquid metal containing a known concentration of oxygen, the saturation solubility at the temperature of interest is required. It is the oxygen potential that determines the corrosion potential of the liquid metal.

Since oxygen solubility in liquid metals is temperature dependent, solubility data are of great technological importance. For example, sodium oxide may precipitate in the cooler parts of a sodium circuit unless care is taken to see that the solubility limits are not exceeded at any point. The low oxygen solubility just above the melting point of sodium is taken advantage of in the "cold trapping" method of sodium purification that is used in most sodium loops including those in LMFBRS.

Before proceeding to discuss the oxygen solubility data for different alkali metals, it may be appropriate to consider the methods for measuring oxygen concentration.

5.3. Measurement of Oxygen Concentration

In the development of analytical methods for the determination of dissolved oxygen in alkali metals, sodium has received the most attention. Therefore, in this section we will first describe briefly the more important methods developed for analyzing sodium for oxygen impurity, and then consider their applicability to other alkali metals. Generally, these methods can be divided into two categories: (1) laboratory techniques which make use of sodium samples delivered to the laboratory and (2) on-line techniques, where the measurement is made directly in flowing sodium. However, we will first discuss the various methods and then consider which of them are amenable to adaptation as on-line techniques.

5.3.1. Amalgamation Method

The amalgamation method for determining oxygen in sodium was developed by Pepkowitz and Judd.[7] This method depends on the physical separation of sodium from sodium oxide by repeated extractions with triple distilled mercury. The sodium oxide, being insolute in the amalgam, floats on the surface. Following the extraction the residue is dissolved in water and the sodium oxide determined either by titration against standard acid or by flame photometry. The size of the sodium sample is determined by titrimetry. From these two determinations, the concentration of oxygen in the sample can be calculated. The original procedure of Pepkowitz and Judd has been improved upon by several workers.[8–10]

In the actual procedure about 20 ml of mercury are introduced into an extractor under an inert atmosphere. The sample is dropped into the extraction vessel. During the amalgamation the temperature is kept down by forced cooling of the extractor. When cool, the amalgam is withdrawn from the bottom leaving the top 1 or 2 ml with the residue floating on it. About ten cycles of extraction are carried out to remove all sodium. The residue is dissolved in distilled water and titrated against 0.005 N HCl to a phenolphthalein end point in order to determine the hydroxyl ion equivalent of the sodium monoxide in the samples. Flame photometry can also be used advantageously for determining the sodium in the residue. To determine the sample size, an excess of standard hydrochloric acid is added to the separated sodium amalgam with vigorous stirring and the solution is back-titrated with standard NaOH to a phenolphthalein end point.

While this method uses simple equipment, considerable care and experience is necessary to get good results. When adequate precautions are taken to ensure the purity of the blanket gas and to remove the last traces of moisture from the extractor, the analytical blank can be brought down to a few tens of micrograms of oxygen. Thus the method is considered reliable for oxygen concentrations greater than 10 ppm. The method was widely used with samples from flow-through samplers. However, it is rather cumbersome and is not considered very reliable for the oxygen levels that are encountered in present-day LMFBR loops. Further, it is not specific as Na_2O may not be the only constituent of the insoluble residue collected by amalgamation. For example, NaH, Na_2CO_3, NaOH, Na_xC_y, etc., may contribute to the measured value.[11]

5.3.2. Distillation Method

In the distillation method, the metallic sodium is preferentially distilled off under vacuum at about 350°C, leaving behind the nonvolatile residue which is mainly sodium monoxide. The residue is dissolved and the sodium oxide is determined either titrimetrically or by flame photometry.

The distillation method was first reported by Humphreys[12] and later improved upon by others.[13–16] Both resistance heating and induction heating have been used for distillation, but the latter has been found to be more convenient. It is important that the temperature of the sample at the end of the distillation should not exceed 400°C to avoid decomposition of Na_2O. A suitable end-point detection is often incorporated in the distillation apparatus.

The distillation apparatus in routine use in the authors' laboratories is shown in Fig. 5.2. The sodium sample contained in a nickel cup is mounted inside the glass distillation apparatus, the whole operation being carried out in an inert-atmosphere glove box. The apparatus is taken out of the glove box after closing it leak-tight, placed in position inside the induction coil, and evacuated to a pressure of $\sim 10^{-5}$ torr. The induction coil around the chimney is then energized to selectively distill off the sodium which gets coated on the walls of the chimney. The temperature of the sample is monitored since it should not be allowed to rise beyond 400°C. After the distillation is over, the crucible is taken out and the residue is dissolved in distilled water. Sodium is estimated by atomic absorption spectrophotometry. Before the distillation of the actual sample, the inner surface of the glass chimney is coated with sodium by distilling a dummy sample in the same chimney in order to eliminate contamination from oxygen and moisture adsorbed on the walls of the distillation apparatus. The present authors have found this method to be very sensitive (~ 1 ppm) and accurate (± 0.5 ppm).

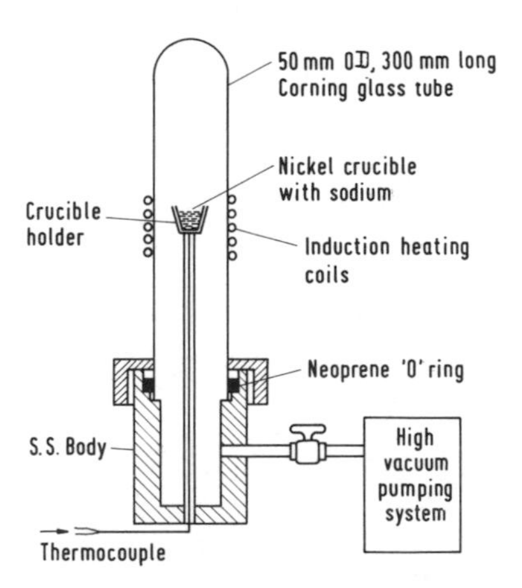

Figure 5.2. Apparatus for analytical sodium distillation.

5.3.3. Electrochemical Oxygen Meter

The galvanic cell method of monitoring oxygen impurity in liquid sodium was first proposed by Horsley.[17] It makes use of the electrochemical cell

$$\text{Na}, [\text{O}]_{\text{Na}}\, |\text{Oxygen-ion-conducting solid electrolyte}|\ \text{O}_2 \text{ reference electrode}$$

Liquid sodium containing dissolved oxygen forms one electrode and an oxygen reference electrode, usually a metal/metal oxide system, the other. They are separated by a solid electrolyte which conducts oxygen ions. If the oxygen potential of the two electrodes are μ_1 and μ_2, the EMF of the cell is given by

$$E = \frac{1}{4F} \int_{\mu_1}^{\mu_2} t_{\text{ion}}\, d\mu_{O_2} \tag{16}$$

where t_{ion} is the ionic transport number for oxygen in the electrolyte and F is the Faraday constant. When $t_{\text{ion}} = 1$, the reversible EMF of the cell is given by

$$E = \frac{RT}{4F} \ln \left[\frac{p_{O_2}^{\text{ref}}}{p_{O_2}^{\text{Na}}} \right] \tag{17}$$

where $p_{O_2}^{\text{ref}}$ and $p_{O_2}^{\text{Na}}$ are the partial pressures of oxygen corresponding to the reference electrode and sodium, respectively.

The solid electrolyte chosen for this application must have low electronic and hole conductivities under the conditions of application ($T = 500$ to 600°C, oxygen concentration at ppm levels) so that the ionic transport number is $\geqslant 0.99$. In addition, the electrolyte must be compatible with sodium. Thoria (ThO_2) doped with 7.5 mol% of yttria (Y_2O_3) has been found to be the best electrolyte for this purpose.[18] The reference electrodes commonly used are of two types:

1. Metal/metal oxide electrodes[19–21] such as In/In_2O_3, Cu/Cu_2O, and Sn/SnO_2;
2. Pt/air reference electrode.[22,23]

The latter electrode, while convenient, gives slightly less than theoretical EMF, on account of the fact that the yttria-doped thoria (YDT) electrolyte has to operate at oxygen partial pressures above its electrolytic domain boundary, whereas the former has been shown to give theoretical EMF.[20]

Two types of oxygen probes are in use in different laboratories. One makes use of a long (15 cm) tube of yttria-doped thoria electrolyte which dips into the sodium, the reference electrode being contained inside the tube. Examples are the Westinghouse[22] and Harwell[21] oxygen meters. This type of oxygen meter is illustrated in Fig. 5.3. The cooling fins ensure that a "freeze seal" is formed, thus protecting the upper portion from sodium vapor and high temperatures. Therefore, O-ring seals can be used at the upper end for leak-tightness. A molybdenum or tungsten wire coming out from the reference electrode is taken out through an appropriate electrical feedthrough which is also leak-tight. The EMF is measured between the electrode and the casing using a high-impedance electrometer.

The second type makes use of a short cup or pellet of YDT brazed to a low-expansion metal tube. Meters of this type have been made by the Central Institute for Nuclear Research (ZfK), GDR,[23] INTERATOM,[24] and General Electric.[20] The GE meter, which is commercially available, is schematically shown in Fig. 5.4. The metal tube carrying the YDT cup fits inside a metal housing through which the flowing sodium circulates. Here the ceramic probe is fully immersed in sodium and hence does not see any thermal gradient. Thermal shocks and thermal gradients are believed to be the major causes of the failure of YDT tubes.

Solid-electrolyte-based oxygen meters can be considered as well established for monitoring oxygen concentrations in the range of a fraction of a ppm to several tens of ppm. The upper limit arises from considerations of meter life, as the formation of ternary oxides can damage the electrolyte.

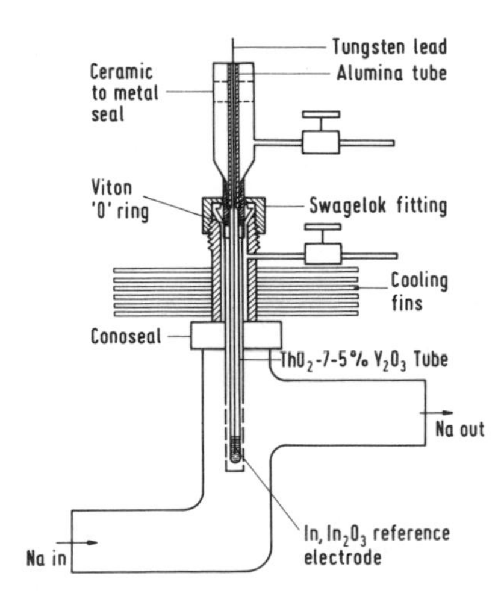

Figure 5.3. Schematic of an oxygen meter using a long electrolyte tube.

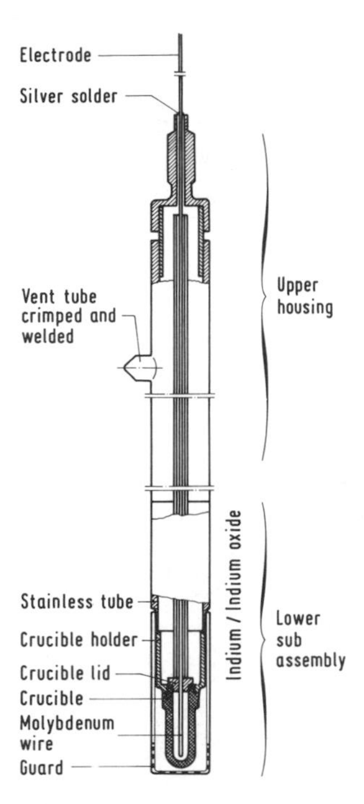

Figure 5.4. Schematic of an electrochemical oxygen meter (General Electric Co., San Jose, Calif., USA, courtesy Dr. P. Roy).

Oxygen meters are commercially available, for example, from the General Electric Company. Meter lifetimes in excess of 10,000 hours have been reported,(25) though shorter lifetimes are not uncommon. In fact, it is difficult to predict the lifetime of a given probe. Calibration of the oxygen probes does pose problems. Cold-trap temperatures,(20) vanadium-wire equilibration,(26) and oxygen potential of a Na–Cr–$NaCrO_2$ system(27) have been used for calibration.

5.3.4. Equilibration Method

Partitioning of oxygen between a refractory metal and a liquid alkali metal can be made use of in the determination of oxygen in the latter. The oxygen distribution coefficient (K_D), which is the ratio of the oxygen concentration in the solid refractory metal(s) to that in the liquid alkali metal (L), is first established as a function of temperature:

$$K_D = \frac{\mathrm{C}_{i,\mathrm{s}}}{\mathrm{C}_{i,\mathrm{L}}} = \frac{C^0_{i,\mathrm{s}}}{C^0_{i,\mathrm{L}}} \exp\left[\frac{\Delta_f G^0(\mathrm{L}) - \Delta_f G^0(\mathrm{s})}{RT}\right] \tag{18}$$

where $C_{i,s}$ and $C_{i,L}$ are the concentrations of oxygen in the solid and liquid, respectively and the zero superscript represents saturation concentration. $\Delta_f G^0(L)$ and $\Delta_f G^0(s)$ are the free energies of formation of the lowest stable oxide in the liquid and the solid, respectively.

The above equation implies the following assumptions:

1. the solid and liquid are mutually insoluble;
2. oxygen levels in the two phases are below saturation solubility; and
3. the dilute solutions of oxygen in the liquid metal and in the refractory metal obey Henry's law.

To measure the oxygen concentration in the alkali metal, a piece of the refractory metal is equilibrated with it at a suitable temperature and the concentration of oxygen in the solid metal is determined. From the known values of K_D and $C_{i,s}$, the oxygen concentration in the alkali metal ($C_{i,L}$) is deduced. The refractory metal can be so chosen that $C_{i,s} \gg C_{i,L}$, and it can be analyzed for oxygen using inert gas fusion techniques. For this method to yield reliable values the following conditions must be satisfied:

(a) The distribution of oxygen between the two metals must reach equilibrium;
(b) The refractory metal must have negligible solubility in the liquid metal;
(c) Oxygen must be in solution in the refractory metal and there should be no compound formation under the conditions of measurement;
(d) The getter metal should not spall or crumble during equilibration.

Smith[28,29] developed a vanadium-wire equilibration technique to measure low concentrations of oxygen in liquid sodium. A 0.25-cm diameter vanadium wire is exposed to liquid sodium until there is equilibrium between the dissolved oxygen in the two phases. The minimum time required for achieving equilibrium is limited by the diffusion of oxygen in the vanadium wire. Smith found that at 750°C the time required was four hours. When the vanadium wire is exposed to sodium in a dynamic loop, a flow rate of $50\ cm^3/min$ is considered sufficient. The exposed vanadium is quickly withdrawn from the sodium, washed with ethanol, and electropolished to remove a 0.01-mm layer which takes care of all surface contamination (20% H_2SO_4 in CH_3OH, Pt cathode, 7V, ~20 s). A portion of the cleaned wire is analyzed for oxygen by inert gas fusion or vacuum fusion techniques.

The following expression for the distribution coefficient of oxygen between vanadium and sodium at 750°C given by Smith can now be used to obtain the oxygen concentration in sodium:

$$\ln K_A = 19.14 + 29.39(1 - N_{0,\mathrm{V}})^2 \tag{19}$$

where $K_A = N_{0,\mathrm{V}}/N_{0,\mathrm{Na}}$ and the N's are the atom fractions of oxygen in vanadium and sodium.

The vanadium-wire equilibration method is considered applicable at low oxygen levels in sodium, and an accuracy of $\pm 15\%$ has been claimed. However, while Smith claims the method to be applicable up to 20 ppm of oxygen in sodium, other authors[30,31] suggest that the limit is much lower. Hooper and Trevillion[30] have concluded after a careful analysis of the vanadium–oxygen system that the formation of β-vanadium limits the applicability of this technique to oxygen levels less than 2 ppm. Lindemer *et al.*[3] suggest that the limit is still lower. However, the success of the method in measuring oxygen in sodium loops and the recent study by nuclear microprobe of vanadium wires exposed to sodium, which established the absence of any radial gradient of oxygen in the wire or precipitation of oxide phases, support the validity of the method up to a few ppm of oxygen in sodium.

5.3.5. Plugging Indicator

The plugging indicator is a simple device that is popularly used in sodium loops.[33,34] It functions as an impurity monitor by indicating the temperature at which the saturable impurity starts precipitating as the sodium stream is slowly cooled. Knowing the saturation solubility of the impurity at the plugging temperature, one obtains the concentration of the impurity in the sodium. Under conditions where oxygen is the only impurity that precipitates on cooling, the plugging indicator becomes an oxygen meter.

A typical plugging indicator is schematically shown in Fig. 5.5. The heart of the device is a cooled orifice or flow restriction. A serrated valve or a perforated plate can also be used to restrict flow. A thermocouple to measure the temperature at the flow restriction, a cooling arrangement, and a flow meter are the other essential components of the device. The cooler is used to slowly reduce the temperature at the orifice. There is a sharp break in flow rate when the impurity starts precipitating. The temperature at which this happens is the plugging temperature. Oxygen solubility at the plugging temperature is taken as the oxygen concentration in the sodium in the loop. When the cooling is stopped, the temperature at

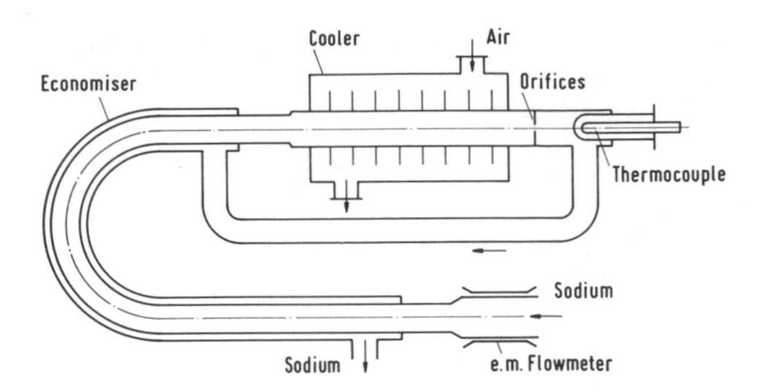

Figure 5.5. Schematic of a plugging indicator.

the orifice returns to normal; the oxide redissolves, thus restoring normal flow.

The chief merit of the plugging indicator is that it is simple and inexpensive. It can be automated to periodically carry out a full cycle of measurement and to record the plugging temperature. Its main disadvantage is that it is not specific for any impurity. Both oxygen and hydrogen respond to the plugging indicator. Some uncertainty also arises from the fact that the plugging temperature may not be exactly the same as the temperature of saturation solubility for the dissolved impurity.

Continuous measurement of the plugging temperature can be realized by operating the plugging meter with a partially blocked orifice so that there is equilibrium between the dissolution and precipitation processes. The plug is then in equilibrium with the solution and the plugging temperature is nearer to the saturation temperature. The possibility of determining the presence and concentration of more than one impurity by the appearance of multiple breaks in sodium flow traces of plugging meters has been suggested by several authors.[35,36]

5.3.6. Activation Analysis

Both neutron activation and photoactivation of oxygen in alkali metals have been used for concentration measurement. Neutron activation analysis employs fast neutrons and makes use of the following nuclear reaction:

$$^{16}O + n \rightarrow {}^{16}N + p - 9.6\ \text{MeV} \tag{20}$$

The short-lived (7.1-s half-life) ^{16}N is estimated by gamma spectrometry, making use of its high-energy gamma radiation. Oxygen in lithium has been determined by this method.[37,38]

Photoactivation analysis is based on the (γ, n) reaction:

$$^{16}O + \gamma \rightarrow {}^{15}O + n \tag{21}$$

involving high-energy gamma rays.[39,40] $^{15}O(t_{1/2} = 2\ \text{min})$ is a pure positron emitter which can be determined with high sensitivity with very little interference from other radioactive isotopes. This technique has been used for the determination of oxygen in sodium by Hislop *et al.*[40]

The major cause of error in the activation analysis method of oxygen determination is the possibility that oxide impurities other than sodium oxide may contribute to the measured value. The best available capsule material that minimizes extraneous oxygen is oxygen-free high-conductivity (OFHC) copper. A sensitivity of a few ppm and a precision of ± 1 ppm appear possible.

5.3.7. Other Methods

A number of other methods have been proposed or developed for measuring oxygen in sodium. In the alkyl halide method,[41] sodium is reacted with an excess of 1-bromobutane in hexane and the sodium monoxide remaining unreacted is determined by titration after the addition of water. This method is, however, no longer used, having been superseded by other methods described above.

Methods based on the measurement of several physical properties have been explored. The use of a rhometer, a device which correlates impurity concentration with electrical resistivity changes, has been investigated.[42] Other examples are ellipsometry and surface tension measurements.[43] These methods have been either not been fully developed or have been found to have many disadvantages.

5.3.8. On-Line Measurements

It is often advantageous to use an on-line meter to continuously monitor the oxygen levels in sodium loops. Apart from the fact that information is instantaneously available and that changes in oxygen levels are recorded, the meter also obviates the need to take out samples, thus minimizing the errors associated with sample handling.

Plugging indicators are the most commonly used on-line meters, but they suffer from several disadvantages such as nonspecificity. Electrochemical oxygen meters are currently being used on many sodium loops. They are specific for oxygen and respond instantaneously to changes in oxygen levels. Vanadium-wire equilibration can be carried out on-line, but the results would be available only after the subsequent chemical analysis. Vacuum distillation can also be performed on-line, but here again this cannot be considered an on-line method as the analysis is carried out off-line.

5.3.9. Applicability of These Methods to Other Alkali Metals

In general, the methods developed for the analysis of sodium may be extended to other alkali metals. One must, however, keep in mind the differences in properties among the alkali metals. The distillation method has been used to measure oxygen solubility in potassium, but not in other alkali metals. The distillation has to be carried out at rather high temperatures in the case of lithium, while in the case of rubidium and cesium lower temperatures are required to prevent the decomposition of the oxide. It must also be kept in mind that oxygen solubility in the higher alkali metals is very high.

The equilibration method can also be developed for other alkali metals, provided appropriate refractory metals and equilibrium conditions are established. Based on the large value of the distribution coefficient of oxygen between zirconium and potassium, an analytical technique known as getter vacuum fusion has been developed.[44] Considering the high solubility of oxygen in Rb and Cs and the temperature at which equilibration is possible with refractory metals, the equilibration method is not attractive for these higher alkali metals.

Electrochemical meters based on YDT can be used for other alkali metals besides sodium. It has, in fact, been used to measure oxygen potentials in potassium[45] and cesium.[5]

All three reported measurements of oxygen in lithium have involved neutron activation analysis.[37,38,46]

5.4. Solubility Data

Solubility of oxygen has been determined in all alkali metals except rubidium. Predictably, sodium has received the lion's share of the attention. The early workers, who carried out oxygen solubility measurements in the fifties, used the amalgamation method. The distillation method was made use of by several investigators who reported oxygen solubility data for sodium in the sixties. More recently the EMF technique and the vanadium-wire equilibration technique have been employed for this purpose. All the relevant details including references are available in the critical reviews by Claxton,[11] Eichelberger,[47] Smith,[48,49] and Noden.[50] Each of the above reviewers has recommended a solubility equation for oxygen in sodium.

Noden, in the latest of the above reviews, has considered the largest number of measurements, and has rejected only the low-temperature data ($<250°C$) based on the amalgamation technique (as the blank error is too large). The equation

$$\log[\mathrm{O}](\mathrm{wppm}) = 6.2571 - \frac{2444.5}{T(\mathrm{K})} \quad (22)$$

recommended by Noden for the maximum solubility by weight of oxygen in sodium in the temperature range 387–828 K may be considered the most acceptable solubility equation. It is also in close agreement with Eichelberger's equation

$$\log[\mathrm{O}](\mathrm{wppm}) = 6.239 - \frac{2447}{T(\mathrm{K})} \quad (23)$$

At 120°C the solubility is 1.2 ppm and this is the basis of sodium purification by cold trapping.

There are only three determinations of oxygen solubility in lithium reported in the literature—those by Hoffman,[37] Konovalov *et al.*,[46] and Yonco *et al.*[38] In all three cases, fast neutron activation analysis on filtered lithium has been used as the technique of measurement. Solubility values have decreased with each measurement and this is consistent with the use of finer filters by the later experimenters. Thus the data of Yonco *et al.*[38] covering the temperature range of 468 to 1007 K may be taken as yielding the best available solubility equation for lithium. Their equation is given below:

$$\log[\mathrm{O}](\mathrm{wppm}) = 6.992 - \frac{2896}{T(\mathrm{K})} \quad (24)$$

At 200°C the solubility is 7 wppm oxygen, thus making it possible to purify lithium by cold trapping to this purity.

Williams[51] measured the solubility of oxygen in liquid potassium in the temperature range of 338 to 783 K using the distillation method. Sreedharan and Gnanamoorthy[52] have derived the following expression from Williams' data:

$$\log[\mathrm{O}](\mathrm{wppm}) = 5.3015 - \frac{876}{T(\mathrm{K})} \quad (25)$$

Oxygen solubility in potassium has also been measured by Ganesan *et al.*[53] in the temperature range of 343 to 473 K. These results yield

Table 5.2. Oxygen Solubility in Alkali Metals (wppm)

Alkali metal	Temperature				
	Near m.p.[a]	473 K	673 K	773 K	873 K
Li	—	7.4	489.6	1763	4735
Na	1.1 (393 K)	12.3	422.3	1246	2867
K	77.6 (350 K)	291.6	893.2	1258.1	1638.4
Cs	26,000 (301.5 K)				

[a] At the specified temperature (given in parentheses) just above the melting point.

solubilities which are lower than Williams' data by at least an order of magnitude. Use of glassware in the experiments and the possible presence of hydroxide contamination in the potassium could have led to Williams' higher values. The results obtained by Ganesan *et al.* are given below:

$$\log[\mathrm{O}](\mathrm{wppm}) = 4.1012 - \frac{774.3}{T(\mathrm{K})} \tag{26}$$

These values are in good agreement with the low-temperature results obtained by Krishnamurthy *et al.*[(54)] and are also consistent with EMF data. Therefore, these results are considered to be more reliable. Even at 350 K this equation yields an oxygen solubility of 77 ppm in potassium. Thus cold trapping has only limited application in the purification of potassium from oxygen.

Oxygen solubility in rubidium and cesium is very high. Knights and Phillips[(5)] have reported the Cs–O phase diagram and oxygen potentials.

Oxygen solubilities in different alkali metals are compared at a few selected temperatures in Table 5.2.

5.5. Oxygen Potentials

Making use of the solubility data, one can calculate the oxygen potentials of liquid alkali metals as a function of oxygen concentration. Here Henry's law is assumed to be valid and this appears to be justifiable on the basis of several measurements of oxygen potentials using electrochemical cells. The other bit of data required for arriving at the oxygen potential expression is the free energy of formation of the monoxide. The oxygen potential expressions given below are taken from the recent review by Lindemer *et al.*[3]

Lithium:

$$\overline{\Delta G}_{O_2}(\text{J/mol}) = -1{,}086{,}500 - 22.32T + 38.29T\log[\text{O}](\text{wppm})$$

Sodium:

$$\overline{\Delta G}_{O_2}(\text{J/mol}) = -740{,}600 + 21.02T + 38.29T\log[\text{O}](\text{wppm})$$

Potassium:

$$\overline{\Delta G}_{O_2}(\text{J/mol}) = -691{,}131 + 117.74T + 38.29T\log[\text{O}](\text{wppm})$$

Cesium:

$$\overline{\Delta G}_{O_2}(\text{J/mol}) = -583{,}800 + 156T + 38.29T\left(\frac{2140}{T} - \frac{1}{3}\right) \times \log\frac{[\text{O}](\text{wppm})}{120{,}300 + 0.8798[\text{O}](\text{wppm})}$$

The oxygen potential expression for cesium is taken from the direct measurement of Knights and Phillips.[5]

The oxygen potentials of alkali metals as a function of oxygen concentration and temperature are plotted in Figs. 5.6. In general, oxygen solubility and oxygen potential increase with the atomic number of the alkali metal. In fact, the solubility is very high in cesium. Though no data are available for rubidium, from the general trend one can place it between potassium and cesium. Solubility of oxygen in rubidium and the oxygen potential of rubidium containing dissolved oxygen may be expected to be high in comparison with the lighter alkali metals.

5.6. Ternary Systems

Dissolved oxygen plays a major role in determining the compatibility of liquid metals with structural and other materials with which they are likely to come in contact in the context of their technological applications. Ternary compounds such as $NaCrO_2$ and Na_4FeO_3 are known to influence the corrosion of steels by sodium containing dissolved oxygen.[55] While Cr_2O_3 does not form at any oxygen level in sodium at the operating temperature of sodium systems, $NaCrO_2$ is formed at ppm levels of oxygen. Such ternary compounds are formed when oxygen concentration exceeds a threshold level in a sodium–steel system. Oxygen in sodium is known to influence the corrosion of steels even below this threshold level required for

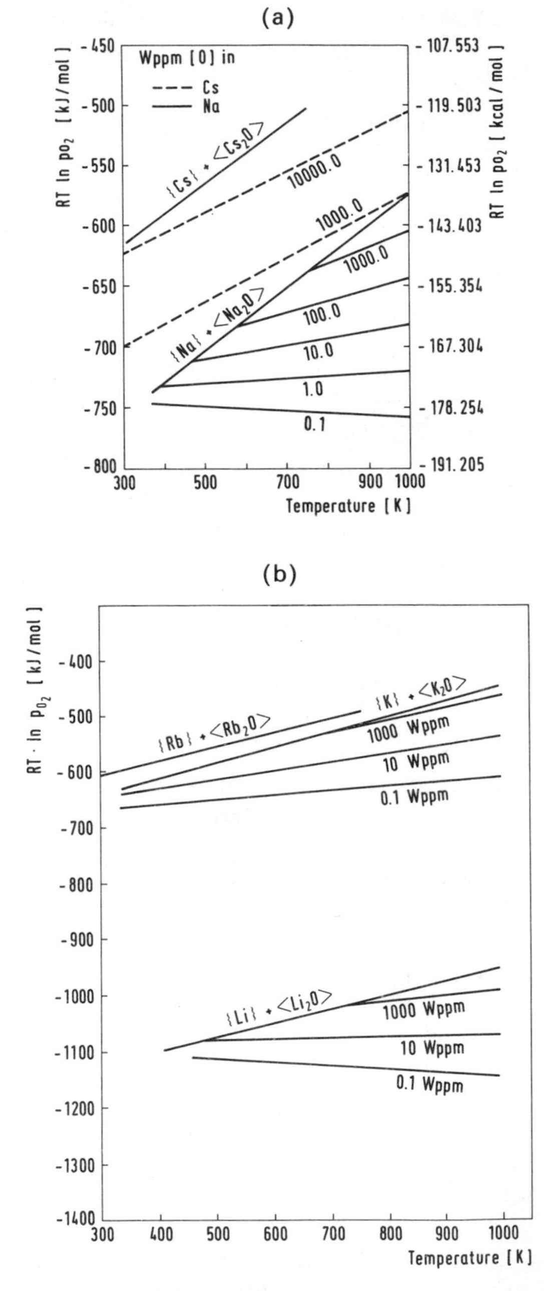

Figure 5.6. (a) Oxygen potentials of sodium–oxygen and cesium–oxygen systems. (b) Oxygen potentials of lithium–oxygen, potassium–oxygen, and rubidium–oxygen systems.

the formation of the ternary compound. Transitory formation of the ternary compound is invoked by Weeks and Isaacs[56] to explain the dependence of corrosion rate on oxygen concentration. Thus, in order to understand corrosion and other chemical reactions between liquid alkali metals and solid structural materials, fuels, etc., it is necessary to understand the chemistry of the ternary system: liquid alkali metal (A)–metal constituent of the solid phase (M)–oxygen. In this section, we consider such ternary systems. As stainless steels and ferritic steels are the main structural materials in the heat transport systems based on alkali metals, the following ternary systems are relevant: A–Cr–O, A–Fe–O, A–Ni–O, A–Mo–O, and A–Mn–O.

Refractory metals/alloys involving tungsten, vanadium, and niobium are likely to be used in controlled thermonuclear reactors (CTRs) and have been used in clads/electrodes. In addition, vanadium is the getter material used in the determination of oxygen concentration in sodium by the equilibration method. Therefore, the ternary systems A–V–O, A–Nb–O, and A–W–O are also considered. As uranium and plutonium are the main constituents of fast-reactor fuels, A–U–O and A–Pu–O systems assume significance. Fission products which have high abundance or which play a major role in understanding fuel behavior are also considered in the context of their interaction with alkali metals. These include rare earths, molybdenum, and zirconium. Cesium and rubidium are themselves fission products. In alkali metal batteries, ceramics such as β-alumina are used so the A–Al–O system is also briefly considered in this section. The references to the compounds mentioned in the following discussion may be found in the excellent review by Lindemer, Besmann, and Johnson[3] unless otherwise indicated.

5.6.1. A–Cr–O System

Alkali metals and chromium form ternary compounds having the formulas $ACrO_2$, $A_2Cr_2O_7$, A_3CrO_4, A_4CrO_4, and A_2CrO_4. $ACrO_2$ is known for all alkali metals except cesium. On the other hand, lithium does not appear to form A_4CrO_4 whereas this compound has been reported in the case of Na, K, and Cs.

In the ternary systems involving lithium and sodium, it is only $ACrO_2$ which exists in equilibrium with one or both the metallic phases. This compound falls in the A_2O–Cr_2O_3 pseudo-binary system whereas all the other compounds are on the oxygen-rich side of this binary line in the ternary phase diagram. In the K–Cr–O and Rb–Cr–O systems, $ACrO_2$ exists in equilibrium with the metallic phases but with increasing oxygen content the A_4CrO_4 phase appears.[57,58] Thermodynamic data on these compounds have been derived from EMF data.[57-59] Rubidium is shown to form a metastable phase of composition Rb_3CrO_4. In the Cs–Cr–O system it is the compound Cs_4CrO_4 which exists in equilibrium with Cs and Cr, and all the other compounds are on the oxygen-rich side of the Cs_4CrO_4–Cr_2O_3 line. The appropriate phase diagrams are given in Fig. 5.7. It may be noted that while $NaCrO_2$ exists in equilibrium with Na and Cr, $LiCrO_2$ exists in equilibrium with Li_2O and Cr. In other words, $LiCrO_2$ is not present with liquid lithium—a consequence of the much higher stability of Li_2O compared to Na_2O. The standard free energies of formation of $LiCrO_2$, $NaCrO_2$, $KCrO_2$, and Cs_4CrO_4 are given in Table 5.3. The threshold oxygen level in sodium above which $NaCrO_2$ forms has been

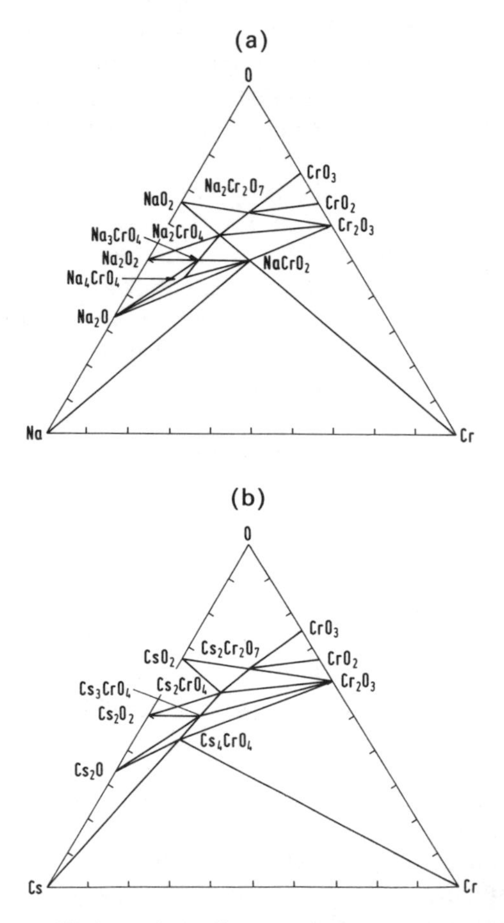

Figure 5.7. (a) The equilibrium phase diagram of the Na–Cr–O system at $T < 633$ K. (b) The equilibrium phase diagram of the Cs–Cr–O system at $T < 660$ K.

Table 5.3. Standard Gibbs Energy of Formation of Alkali Metal Chromites

Compound	$\Delta_f G^0 = -A + BT$ (J/g-atom O_2)	Temp. range (K)	Reference
$LiCrO_2$	$-1005 \times 10^3 + (256.2)T$	690–1000	56
$NaCrO_2$	$-876 \times 10^3 + (194)T$	510–1100	59
	$-869.68 \times 10^3 + (185.75)T$	700–1200	57
$KCrO_2$	$-739 \times 10^3 + (102.2)T$	623–773	105
	$-803.84 \times 10^3 + (152.53)T$	700–1200	106
Cs_4CrO_4	$-1480 \times 10^3 + (340)T$	600–900	59

directly measured in the Na–Cr–$NaCrO_2$ phase field by EMF techniques.[60,61] This threshold can also be calculated from the free energy data on $NaCrO_2$ which have been obtained by experimental measurements in other phase fields where $NaCrO_2$ also exists.[5,62–65] These results are given in Table 5.4. It can be seen that there is considerable disparity among the various values of threshold oxygen concentrations. Direct in-sodium measurements have yielded higher values whereas calculations from free energy data provide lower values. This system obviously needs more careful examination.

5.6.2. A–Fe–O System

The compounds in this system may be considered as formed from A_2O, on the one hand, and FeO and Fe_2O_3, on the other. Clearly, the compounds in the A_2O–FeO pseudo-binary line are more important as they are likely to exist in equilibrium with at least one of the metals. There are two such compounds, Na_4FeO_3 and Na_2FeO_2, in the sodium system. They are not established in the case of the other alkali metals. An equilibrium diagram based on the estimated thermodynamic properties of the ternary compounds has been given by Lindemer *et al.*[3] for the system in the temperature range 720–880 K. According to this, Na_4FeO_3 forms a ternary phase field with Fe and Na. Dai *et al.*[66] have recently studied the phase relationships in this system. They have not observed phases such as $Na_4Fe_6O_{11}$; instead they have found that Na_2O dissolves in wustite up to the saturation limit. This sodium oxide-saturated wustite was found to be in equilibrium with Fe and $NaFeO_2$. From the standard free energy of formation of Na_4FeO_3, estimated by Lindemer *et al.*,[3] one can compute the threshold oxygen concentration in sodium for the formation of sodium ferrite. At 800 K the value is 892 ppm.

In the case of higher alkali metals such as potassium and rubidium,

Table 5.4. Threshold Oxygen Levels in Liquid Sodium for the Formation of $NaCrO_2$

Temperature	Calculated values from free energy data (wppm)				Measured values (wppm)	
(K)	Knights & Phillips (ref. 65)	Shaiu *et al.* (ref. 64)	Bhat *et al.* (ref. 62)	Sreedharan *et al.* (ref. 63)	Adamson (ref. 60)	Periaswami *et al.* (ref. 61)
	$a_{Cr} = 0.29$/18-8 steels				$a_{Cr} = 0.29$/18-8 steels	
700	0.015	0.44	0.436	0.332	6.52	18.62
800	0.65	1.87	2.07	1.21	14.0	23.85
900	2.02	5.72	9.04	3.3	25.4	29.88
1000	5.01	13.98	26	7.34	40.3	35.07
	$a_{Cr} = 1.0$				$a_{Cr} = 1.0$	
700	0.081	0.24	0.105	0.18	3.49	9.89
800	0.35	1.01	0.57	0.652	7.47	12.72
900	1.09	3.08	2.15	1.77	13.55	15.98
1000	2.70	7.53	6.24	3.96	22.08	18.83

compounds of the composition $A_6Fe_2O_6$ exist in equilibrium with both the metallic phases. Thermodynamic data for these compounds have been reported.[57,67]

5.6.3. A–Ni–O and A–Co–O Systems

Several ternary compounds of alkali metals with nickel and oxygen have been reported. Compounds of the formulas $ANiO_2$ (A = Li, Na), A_2NiO_2 (A = Na, K), and A_2NiO_3 (A = Li, K, Rb) are known, but very little thermodynamic information exists. Free energy data for $NaNiO_2$ and Na_2NiO_2 have been obtained by EMF measurements.[64] In general, these compounds are formed at relatively high oxygen potentials and hence are unlikely to be present in equilibrium with the alkali metal. Thus, they are not very important from the point of view of the applications of liquid alkali metals.

In the Na–Co–O system some compounds such as Na_4CoO_3, $Na_4Co_4O_9$, $Na_{10}Co_4O_9$, and $NaCoO_2$ have been reported.[68,69] However, thermodynamic data and phase relationships are not available.

5.6.4. A–V–O System

Vanadium forms a large number of oxides: V_2O_5, VO_2, V_2O_3, VO, V_2O, V_5O, and V_9O, the last four showing ranges of stoichiometry. Several of these form double oxides with alkali metal oxides. Thus ternary compounds with the formulas A_3VO_4, $A_4V_2O_7$, AVO_3, etc., are known. AVO_2 is reported for A = Li, Na, and K. Since vanadium exists in its lowest oxidation state in this compound, this is more likely to figure in alkali metal systems than the higher oxide compounds. The equilibrium diagram of the Na–V–O ternary system (Fig. 5.8) shows that $NaVO_2$ indeed exists in equilibrium with liquid sodium and V_5O.[3] It may be noted that V_9O is not shown, as it is stable only up to 783 K. When the oxygen level in sodium in contact with vanadium ($T = 900$ K) is very low, α-vanadium is formed. As the oxygen level increases, V_5O is formed and later a Na–$NaVO_2$–V_5O equilibrium is established. There is considerable confusion in the literature regarding the threshold oxygen level for V_5O formation. Recently the equilibrium oxygen potential in sodium with $NaVO_2$ has been measured in the temperature range 800–950 K.[70] Contrary to the phase relationships derived by Lindemer *et al.*,[3] this work indicated that the coexisting oxide phase is "VO" instead of V_5O. The threshold oxygen concentration in sodium for the formation of $NaVO_2$ as calculated from the measured equilibrium oxygen potential at 1000 K is 36.7 ppm. The oxygen concentration necessary for the formation of the

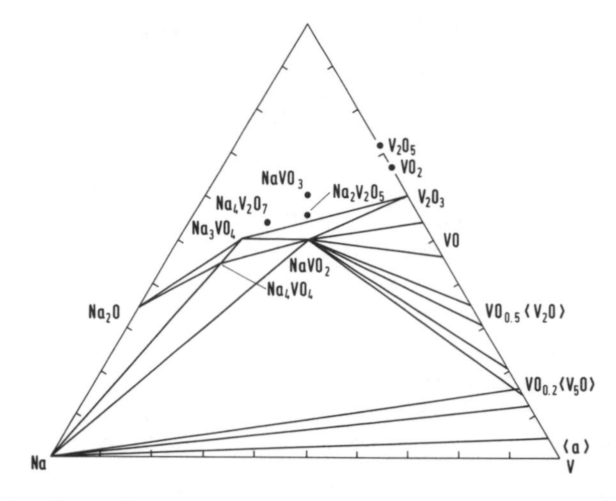

Figure 5.8. The equilibrium phase diagram of the Na–V–O system at $T = 900$ K.

vanadium-rich oxide phase, as estimated by different authors, is given in Table 5.5. In the case of the Li–V–O system, it is Li_2O which is the first phase to be formed as the oxygen level is raised and hence $LiVO_2$ is not considered in detail here.

5.6.5. A–Nb–O System

Niobium forms oxides such as NbO, NbO_2, and Nb_2O_3. No compound is known in the pseudo-binary system Na_2O–NbO. The possible formation of compounds like A_3NbO_4 (known for all alkali metals) and $ANbO_2$ (A = Li, Na) in alkali metals must, therefore, be considered. In lithium, the stability of Li_2O ensures that no ternary compound can coexist in equilibrium with liquid lithium. In liquid sodium Na_3NbO_4 is believed to form the ternary phase field with liquid Na and Nb. Frankham[70] has measured the equilibrium oxygen potential in this phase field, and from

Table 5.5. Threshold Oxygen Concentration for V_5O Formation

Temp. (K)	Estimates of Hooper and Trevillion[a]	Experimental results of Smith[a]
923	0.999	1.785
973	1.411	5.714
1023	1.905	15.502

[a] Data taken from ref. 71.

these data the following expression has been derived for the free energy of formation of Na_3NbO_4:

$$\Delta_f G^0(Na_3NbO_4) = -1{,}504{,}667 + 181.127T \text{ J/mol} \qquad (810\text{–}917 \text{ K}) \quad (27)$$

5.6.6. A–Mo–O System

Molybdenum forms several oxides—MoO_2, MoO_3, Mo_4O_{11}, etc. Thus there is the possibility of forming a host of ternary oxides, many of which have been identified. However, the phase relationships have yet to be established.

In the Li–Mo–O system, there is no double oxide present in equilibrium with the liquid metal,[72] as is to be expected. Very few data are available on other phase fields. Standard heat of formation values are available for Li_4MoO_5 and Li_2MoO_4 from recent Knudsen effusion studies.[74]

In the Na–Mo–O system the reported ternary compound in which Mo exists in its lowest oxidation state is $NaMoO_2$. However, this compound has not been clearly identified in equilibrium with sodium.[72,73] Na_4MoO_5 has been reported to be in equilibrium with sodium and molybdenum at 600°C but the double oxide in equilibrium with the metals at 400°C is Na_2MoO_4.[72] Molybdenum also forms a series of polymolybdates with varying Na_2O/MoO_3 ratios. A phase diagram based on results of a recent work[75] at 923 K is given in Fig. 5.9.

In the K–Mo–O system, K_2MoO_4, K_4MoO_5,[76] and $K_6Mo_2O_9$[77] have so far been identified as compounds in the system apart from bronzes

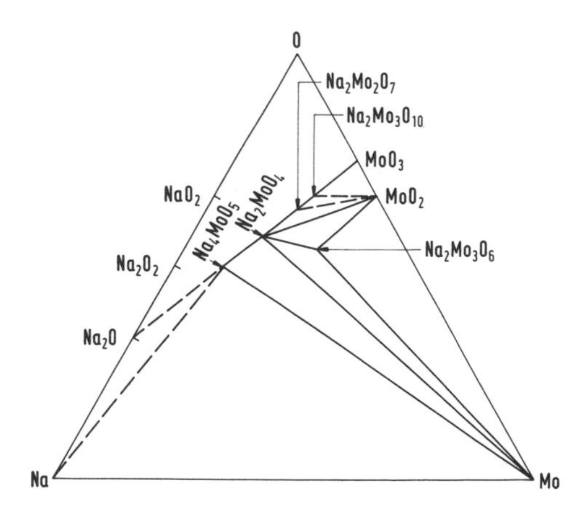

Figure 5.9. The equilibrium phase diagram of the Na–Mo–O system at $T = 923$ K.

and polymolybdates. An equilibrium phase diagram has yet to be arrived at.

The Cs–Mo–O system has a series of mixed oxides/polymolybdates with different Cs_2O/MoO_3 ratios. This system has been dealt with in some detail by Lindemer *et al.*[3] The equilibrium diagram taken from this reference is given in Fig. 5.10.

5.6.7. A–Mn–O System

The ternary systems A–Mn–O have not been intensively studied, although compounds with manganese valency varying from +2 to +7 have been reported. $AMnO_2$, A_2MnO_3, A_3MnO_4, and A_2MnO_4 (A = Li, Na, K) have been identified, apart from the well-known permanganates, $AMnO_4$. In the Na–Mn–O system additional compounds such as $Na_4Mn_2O_5$, $NaMn_7O_{12}$, Na_4MnO_4, and Na_5MnO_5 have been characterized. In the K–Mn–O system K_4MnO_4 is also known. Compounds of the type A_xMnO_2 ($x<1$) have been reported in the case of sodium and potassium.[78,79] MnO is found to be stable in sodium. With increasing oxygen levels in sodium, $NaMnO_2$ and $Na_4Mn_2O_5$ are progressively formed.[80]

5.6.8. A–Al–O System

Alumina ceramics are used for the containment of alkali metals because of their good compatibility. Jung *et al.*[81] and Fink[82] reported that high-purity alumina withstood sodium service with very little damage although surface discoloration was observed. Both workers reported that

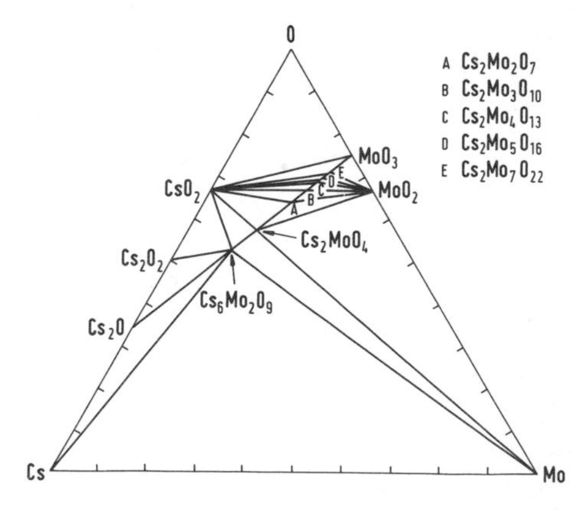

Figure 5.10. The phase diagram of the Cs–Mo–O system at $T < 730$ K (after ref. 3).

the attack on alumina ware was extensive when impurities such as SiO_2 and Cr_2O_3 were present. Jung *et al.* further observed that the presence of β-Al_2O_3 also leads to varying degrees of attack by sodium. Barker *et al.*[83,84] have reported alkali-rich sodium aluminates such as $Na_7Al_3O_8$, Na_5AlO_4, and $Na_{17}Al_5O_{16}$ in their study on the Na_2O–Al_2O_3 system. Measurement of thermodynamic data for $NaAlO_2$ has been carried out recently.[70] However, the phase relationships in this system are not yet clear.

In the Li–Al–O system compounds such as Li_5AlO_4, Li_3AlO_3, $LiAlO_2$, and $LiAl_5O_8$ have been reported but the phase diagram has not yet been established completely. Lithium metal has been found to attack α-Al_2O_3 ceramic at 800 K exothermically. From the products of this reaction Konys *et al.*[85] proposed two new tie lines which encloses the phase field Li_5AlO_4–$LiAlO_2$–Li_9Al_2. Vaporization and thermal stability of lithium aluminates have been studied by Knudsen cell mass spectrometry.[74]

5.6.9. A–W–O System

Several ternary compounds and bronzes are known in A–W–O systems. Li_2WO_4, Li_4WO_5, and Li_6WO_6 are compounds belonging to the Li_2O–WO_3 pseudo-binary system. The existence of $LiWO_2$ has also been reported.[77] Liquid lithium reduces tungsten oxides to metallic tungsten. The phase relationships in this system have not yet been established. Liquid sodium also reduces tungsten oxides to metallic tungsten, and a ternary oxide is formed. At 600°C Na_3WO_4 is formed in low-oxygen sodium whereas Na_6WO_6 is formed when oxygen levels are high.[86] Na_2WO_4 is found to be stable in liquid sodium at 400°C.[72] Potassium forms K_2WO_3, K_4WO_5, and $K_6W_2O_9$ in the potassium-rich regions.[76,77] Bronzes of the formula A_xWO_3 $(0 \leqslant x \leqslant 1)$ are known for all alkali metals except lithium. These bronzes have unusual electrical properties which evoke great interest in them.

5.6.10. A–Zr–O System

In the A–Zr–O systems (A = Li, Na, K) no ternary oxide compound exists in equilibrium with the alkali metal and zirconium.[86,72] Ternary compounds with equimolar ratio of A_2O and ZrO_2 are known for all alkali metals other than rubidium. As expected from free energy data, liquid lithium reduces ZrO_2 to metallic zirconium. However, ZrO_2 is stable towards liquid sodium and potassium. Na_2ZrO_3 forms in sodium with increasing oxygen content. Heat of formation and heat capacity data for Li_2ZrO_3 and Na_2ZrO_3 have been reported.[87,88]

5.6.11. A–U(Pu)–O System

The importance of these ternary systems stems from the fact that uranium dioxide and uranium–plutonium mixed oxide are used as reactor fuels. In LMFBRs the possibility of reaction between the mixed oxide fuel and the liquid sodium coolant must be considered, as a consequence of a breach in fuel cladding. Cesium and rubidium are fission products and hence their reactions with the oxide fuel are of interest in understanding fuel chemistry. The ternary systems involving uranium are much better studied than those involving plutonium. Therefore, we discuss here the uranium systems and refer to any information available on the plutonium systems at the appropriate places.

5.6.11.1. Li–U–O System

An equilibrium phase diagram of this ternary system is not available. There are several compounds known[89,90] in the Li_2O–UO_3 pseudo-binary system: Li_6UO_6, Li_4UO_5, Li_2UO_4, $Li_6U_5O_{18}$, and $Li_2U_3O_{10}$. Lower-valent uranates obtained by the reduction of hexavalent uranates are: Li_7UO_6, Li_3UO_4, $LiUO_3$, and $Li_2U_4O_{11}$. According to Lindemer *et al.*,[3] the compound in equilibrium with liquid lithium is likely to be Li_7UO_6. This compound, along with Li_3UO_4, coexists with UO_2.

5.6.11.2. Na–U–O System

Much attention has been paid to this system on account of the possible harmful effects of fuel–coolant chemical interaction in LMFBRs. In the event of a clad breach in a fuel pin, liquid sodium is likely to enter the pin. The reaction between sodium and the mixed oxide fuel results in the formation of sodium uranoplutonate. This compound is of lower density (5.6 g/cm^3) than the mixed oxide and hence its formation can lead to fuel swelling and failure propagation.

There are several well-characterized compounds in the Na_2O–UO_3 pseudo-binary system: Na_2UO_4, $Na_2U_2O_7$, and Na_4UO_5. In addition, there are two other sodium uranates in which uranium exists in a lower valency, viz., Na_3UO_4 and $NaUO_3$. Of these, the most important compound is Na_3UO_4, as it can coexist in equilibrium with liquid sodium and UO_2. Thus, it is this compound which is formed when sodium comes in contact with the oxide fuel. The equilibrium phase diagram of the ternary system is given in Fig. 5.11.

There have been several investigations of the properties of Na_3UO_4 (as well as $Na_3(U, Pu)O_4$) and the conditions of its formation in liquid sodium.[61,91–94] The free energy of formation of Na_3UO_4 is[95]:

$$\Delta_f G^0(Na_3UO_4) = -2{,}027{,}177 + 432.21T \text{ J/mol} \qquad (28)$$

The oxygen potential of liquid sodium in equilibrium with Na_3UO_4 and UO_2 has been measured by Adamson *et al.*,[92] Periaswami *et al.*,[61] and Mignanelli and Potter.[94] The temperature dependence of the equilibrium oxygen potential can be expressed by the equation[95]:

$$\overline{\Delta G}^{eq}_{O_2} = -944{,}951 + 261.34T \text{ J/mol } O_2 \qquad (29)$$

The reaction of sodium with PuO_2 results in the formation of sodium plutonate, Na_3PuO_4, whose free energy of formation is not established. The oxygen potential of sodium in equilibrium with plutonium oxide and sodium plutonate has been measured by Mignanelli and Potter[96] and their results are comparable within experimental uncertainties to those for the corresponding uranium system.

It is the reaction between liquid sodium and the mixed oxide fuel that is of real interest in fast breeder reactors. The appropriate quaternary phase diagram of the Na–U–Pu–O system is rather complicated, but we are concerned here with only the phase field containing sodium (with oxygen dissolved in it), the mixed oxide, and sodium uranoplutonate. The mixed

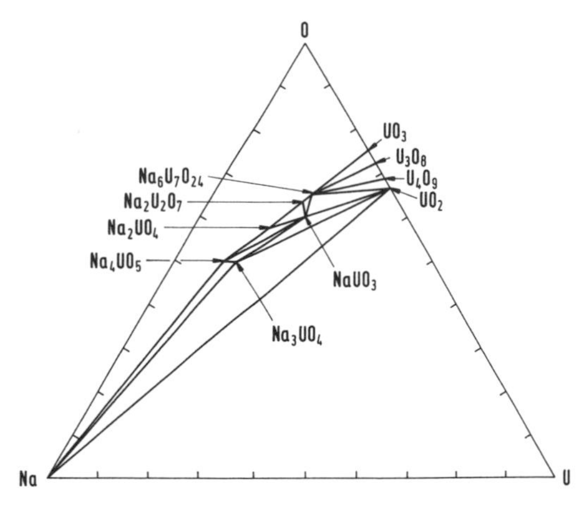

Figure 5.11. The equilibrium phase diagram of the Na–U–O system at $T < 950$ K (after ref. 3).

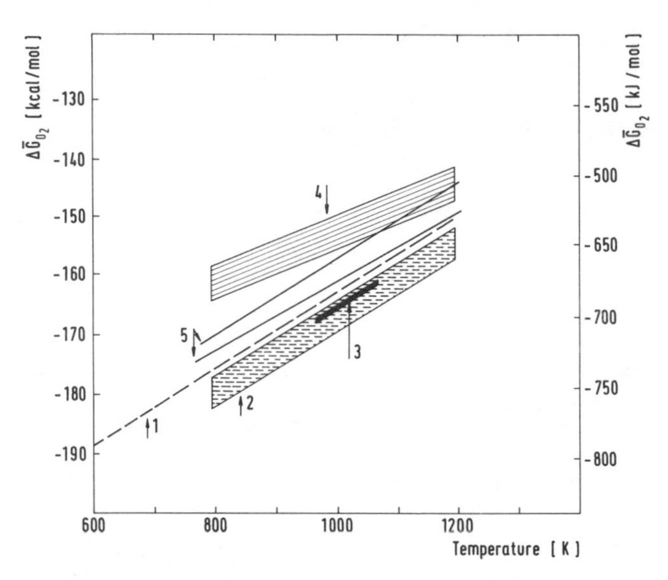

Figure 5.12. Oxygen potentials of mixed oxide fuels and Na–Na_3MO_4 (M = U, Pu)–(U, Pu) O_{2-x} (after refs. 94 and 96). Oxygen potentials in the phase field Na–MO_{2-x}–Na_3MO_4 (M = U, Pu). *1:* Calculated line for M = U; *2, 4, 5:* Calculated values based on the $\overline{\Delta G}_{O_2}$ *vs.* O/M relations given in Woodley (Ref. 111), Markin and McIvor (Ref. 109), and Krishnaiah and Sriramamurthy (Ref. 108), respectively; *3:* Measured values of Woodley and Adamson (Ref. 111) for $M = U_{0.75}Pu_{0.25}$.

oxide would be hypostoichiometric, the actual O/M ratio being related to the prevailing oxygen potential. There have been several studies aimed at measuring the equilibrium oxygen potential of this phase field and characterizing the reaction products, $U_{1-x_1}Pu_{x_1}O_{2-y}(s)$ and $Na_3U_{1-x_2}Pu_{x_2}O_4(s)$.[92,94,97] Here the plutonium fraction of the heavy metals in the two phases are shown as x_1 and x_2, as they can have different values.* However, recent measurements by Mignanelli and Potter[94,96] indicate that the U/Pu ratio is the same in both these phases. The measured and calculated oxygen potentials of this phase field are given in Fig. 5.12.

5.6.11.3. K–U–O and Rb–U–O Systems

In the K–U–O system a number of uranates have been characterized.[98] These include the hexavalent uranates, K_4UO_5, K_2UO_4, $K_2U_2O_7$, $K_2U_4O_{13}$, and $K_2U_7O_{22}$, which fall on the K_2O–UO_3 line, and

* y is the departure from stoichiometry of the hypostoichiometric mixed oxide.

a pentavalent uranate, KUO_3. Reliable thermodynamic data are, however, available only for K_2UO_4.[99]

Very similar is the case of the Rb–U–O system. The known ternary compounds are: Rb_4UO_5, Rb_2UO_4, $Rb_2U_2O_7$, $Rb_2U_4O_{13}$, $Rb_2U_7O_{22}$, and $RbUO_3$.[98,99] Thermodynamic data are not available for any of these compounds except Rb_2UO_4.

Phase diagrams of these ternary systems are not well established. Hardly any data are available on the corresponding ternary systems involving plutonium.

5.6.11.4. Cs–U–O System

The reaction of the fission product cesium with UO_2 and (U, Pu) O_2 has a strong bearing on the performance of the oxide fuel. Cesium migrates both radially and axially in a fuel pin. Thus in a fast reactor fuel pin a high concentration of cesium develops at the cooler ends, and this leads to a reaction between cesium and the insulator (UO_2) pellet. The product, cesium uranate, is much more voluminous than UO_2 and hence its formation causes fuel pin swelling. The availability of cesium to tie down tellurium, which is believed to cause clad corrosion, is also controlled by the formation of cesium uranate as well as other cesium compounds with chromium or fission products. Thus the Cs–U–O system has been the focus of much attention.

A number of cesium uranates have been identified by different workers.[99–101] Cs_2UO_4, $Cs_2U_2O_7$, $Cs_4U_5O_{17}$, $Cs_2U_4O_{13}$, $Cs_2U_5O_{16}$, $Cs_2U_7O_{22}$, and $Cs_2U_{15}O_{46}$ are compounds belonging to the Cs_2O–UO_3 pseudo-binary system. Uranium exists in a lower valency in the following compounds: $Cs_2UO_{3.56}$, $Cs_2U_4O_{12}$, $Cs_2U_6O_{18}$, and $Cs_2U_9O_{27}$. The phase that exists in equilibrium with liquid cesium and UO_2 is $Cs_2UO_{3.56}$. The relevant portion of the Cs–U–O phase diagram is shown in Fig. 5.13.

The phase diagram of the Cs–Pu–O system is not known. In fact, even the compounds that form when liquid cesium reacts with PuO_2 and (U, Pu) O_2 have not been characterized.

5.6.12. A–Ln–O System

Compounds of the formulas $NaLnO_2$ and Na_2LnO_3 have been reported in the sodium–lanthanide (Ln)–oxygen system.[102,103] These compounds have, in general, been prepared by solid-state reactions. $ALnO_2$, in which the lanthanide is present in the lower valency state, is more likely to

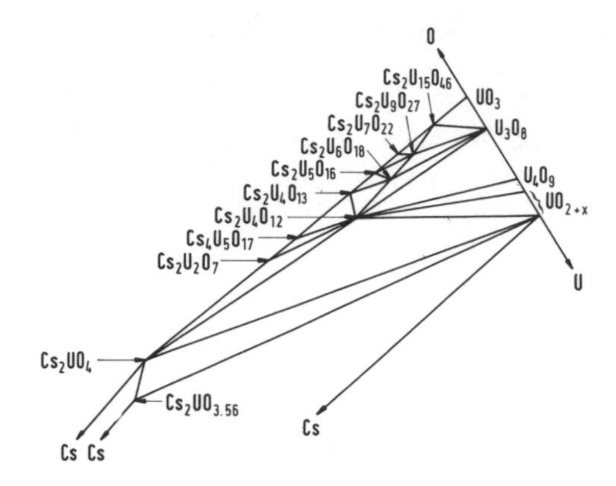

Figure 5.13. Phase diagram of the Cs–U–O system (after ref. 101).

be of interest in the presence of the liquid metal phase. The $NaLnO_2$ compounds with Ln = La, Nd, Sm, Eu, and Gd have been prepared by Blasse[104] using solid-state reactions. $KCeO_2$ was prepared by Clos *et al.*[105] More recently, Mignanelli and Potter[106] investigated the Na–Ce–O system and identified $NaCeO_2$ as existing in equilibrium with sodium. In a comprehensive study Barker *et al.*[107] prepared $NaLnO_2$ with Ln = Pr, Nd, Sm, Eu, Gd, Dy, Ho, and Er by reacting the lanthanide metal or oxide with sodium monoxide or peroxide at around 650°C. They also observed that the ternary orides were formed when the lanthanide oxide reacted with sodium monoxide in liquid sodium. The heavier rare earths gave only the cubic form of the ternary oxide. The lighter rare earths gave the cubic form at low temperatures (<600°C) and a tetragonal form at higher temperatures. The phase diagrams of these ternary systems have not been well established.

References

1. J. C. Bailer, H. J. Emeleus, Sir Ronald Nyholm, and A. F. Trotman-Dickenson (Eds.), *Comprehensive Inorganic Chemistry*, Vol. 1, Pergamon Press, Oxford, 1973.
2. N. G. Vannerberg, in: *Progress in Inorganic Chemistry* (F. A. Cotton, Ed.), Interscience, New York, 1962, Vol. 4, pp. 125–197.
3. T. B. Lindemer, T. M. Besmann, and C. E. Johnson, *J. Nucl. Mater. 100*, 178–226 (1981).
4. R. C. Weast (Ed.), *CRC Handbook of Chemistry and Physics*, Chemical Rubber Co. Press Inc., West Palm Beach, Florida, 1979.
5. C. F. Knights and B. A. Phillips, *J. Nucl. Mater. 84*, 196–206 (1979).
6. Ph. Touzain, *Can. J. Chem. 47*, 2639–2643 (1969).
7. L. P. Pepkowitz and W. C. Judd, *Anal. Chem. 22*, 1283–1286 (1950).
8. D. F. Boltz, J. O. Kermoshchuk, and S. A. Meacham, USAEC Report APDA-168, 1965.

9. J. M. Scarborough and P. F. DeVries, *Anal. Chem. 39*, 826–829 (1967).
10. Ph. Berge, C. Oberlin, and P. Saint Paul, *J. Nucl. Mater. 57*, 283–286 (1975).
11. K. T. Claxton, *J. Nucl. Energy 19*, 849–889 (1965).
12. J. R. Humphreys, Jr., *Sampling and Analysis for Impurities in Liquid Sodium Systems*, Chemical Engineering Progress Symposium Series 53, No. 20, American Institute of Chemical Engineers, New York, 1957, pp. 7–10.
13. K. S. Bergstresser, G. R. Waterbury, and C. F. Metz, USAEC Report LA-3343, 1965.
14. J. A. J. Walker, E. D. France, and W. T. Edwards, *Analyst 90*, 727–731 (1965).
15. M. Hissink, *Atomenergie 12*, 165–170 (1970).
16. H. U. Borgstedt, Z. Peric, A. Marin, and G. Wittig, *Atomwirtschaft-Atomtechnik 17*, 361–362 (1972).
17. G. W. Horsley, UKAEA Report AERE-R-3037, 1961.
18. M. R. Hobdell and C. A. Smith, *J. Nucl. Mater. 110*, 125–139 (1982).
19. B. Minushkin and M. Kolodney, United Nuclear Corporation Report UNC-5131, 1967.
20. P. Roy and B. E. Bugbee, *Nucl. Technol. 39*, 216–218 (1978).
21. R. Thompson, R. G. Taylor, R. C. Asher, C. C. H. Wheatley, and R. Dawson, in: *2nd Intern. Conf. on Liquid Metal Technology in Energy Production* (J. M. Dahlke, Ed.), National Techn. Information Service, Springfield, Va., 1980, (CONF-800401), Papers 16–19.
22. J. M. McKee, D. R. Vissers, P. A. Nelson, B. R. Grundy, E. Berkey, and G. R. Taylor, *Nucl. Technol. 21*, 217–227 (1974).
23. H. Ullmann, T. Reetz, H. Teske, F. A. Kozlov, E. K. Kuznecov, and P. S. Kozub, *Kernenergie 18*, 221–229 (1975); H. Ullmann, K. Teske, and T. Reetz, *Kernenergie 16*, 291–298 (1973).
24. J. Jung, *J. Nucl. Mater. 56*, 213–220 (1975).
25. G. J. Licina, P. Roy, H. Nei, and A. Kakuta, in: *2nd Intern. Conf. on Liquid Metal Technology in Energy Production* (J. M. Dahlke, Ed.), National Techn. Information Service, Springfield, Va., 1980, (CONF-800401), 16–15.
26. J. T. Holmes and G. O. Haroldsen, *Nucl. Technol. 21*, 228–234 (1974).
27. N. P. Bhat and H. U. Borgstedt, *Nucl. Technol. 52*, 153–161 (1981).
28. D. L. Smith, *Nucl. Technol. 11*, 115–119 (1971).
29. D. L. Smith, *J. Less-Common Metals 31*, 345–358 (1973).
30. A. J. Hooper and E. A. Trevillion, *J. Nucl. Mater. 66*, 88–96 (1977).
31. R. F. Keough, *J. Nucl. Mater. 75*, 294–296 (1978).
32. J. W. McMillan, in: *Analysis of Non-Metals in Metals* (G. Kraft, Ed.), de Gruyter, Berlin, 1981, pp. 173–191.
33. I. L. Gray, R. L. Neal, and B. G. Voorhees, *Nucleonics 14*, 34–37 (1956).
34. C. A. Smith, P. A. Smith, and G. Hughes, *Nucl. Energy 18*, 201–214 (1979).
35. A. N. Hamer, J. H. Higson, J. Mathison, and R. Swinhoe, in: *Liquid Alkali Metals*, British Nuclear Energy Society, London, 1973, pp. 59–64.
36. M. Rajan and R. D. Kale, in: *Material Behavior and Physical Chemistry in Liquid Metal Systems* (H. U. Borgstedt, Ed.), Plenum Press, New York, 1982, pp. 81–88.
37. E. E. Hoffman, USAEC Report, ORNL-2674, 1959; ORNL-2894, 1960.
38. R. N. Yonco, V. A. Maroni, J. E. Strain, and J. H. DeVan, *J. Nucl. Mater. 79*, 354–362 (1979).
39. M. Holm and W. M. Sanders, *Nucl. Applications 3*, 308–313 (1967).
40. J. S. Hislop, R. Thompson, and D. A. Wood, in: *Analysis of Non-Metals in Metals* (G. Kraft, Ed.), de Gruyter, Berlin, 1981, pp. 321–330.
41. J. C. White, W. J. Ross, and R. Rowan, Jr., *Anal. Chem. 26*, 210–213 (1954).
42. R. L. Blake and A. R. Eames, *Nucleonics 19*, 66–72 (1961).

43. D. Ensminger, D. R. Greiser, E. H. Hall, J. W. Kissel, I. McCallum, and W. H. Goldthwaite, Report BMI-1409, Batelle Memorial Institute, Columbus, Ohio, 1960, pp. 19–20.
44. R. L. Klueh, *J. Nucl. Energy 25*, 253–261 (1971).
45. B. Minushkin and K. Goldman, in: *Alkali Metals*, The Chemical Society, London, 1967, pp. 403–415.
46. E. E. Konovalov, N. I. Seliverstov, and V. P. Emelyanov, *Izv. Akad. Nauk. SSSR, Metal. 3*, 109–112 (1968).
47. R. L. Eichelberger, USAEC Report, AI-AEC-12685, 1968.
48. D. L. Smith, *Nucl. Technol. 11*, 115–119 (1971).
49. D. L. Smith and R. H. Lee, USAEC Report, ANL-7891, 1972.
50. J. D. Noden, *J. Br. Nucl. Energy Soc. 12*, 57–62, 329–331 (1973).
51. D. D. Williams, *J. Phys. Chem. 63*, 68–71 (1959).
52. O. M. Sreedharan and J. B. Gnanamoorthy, *J. Nucl. Mater. 89*, 113–128 (1980).
53. V. Ganesan, Ch. Adelhelm, and H. U. Borgstedt, *J. Less-Common Metals 113*, 253–259 (1985).
54. D. Krishnamurthy, private communication, 1985.
55. M. G. Barker and D. J. Wood, in: *Chemical Aspects of Corrosion and Mass Transfer in Liquid Sodium* (Sven A. Jansson, Ed.), Metallurgical Society of AIME, New York, 1973, pp. 365–379.
56. J. R. Weeks and H. S. Isaacs, in: *Advances in Corrosion Science and Technology*, Vol. 3, Plenum Press, New York, 1973, pp. 1–66.
57. V. Ganesan and H. U. Borgstedt, *J. Less-Common Metals 114*, 343–354 (1985).
58. P. G. Gadd and H. U. Borgstedt, *J. Nucl. Mater. 119*, 154–161 (1983).
59. O. M. Sreedharan, B. S. Madan, R. Pankajavalli, and J. B. Gnanamoorthy, in: *Liquid Metal Engineering and Technology*, British Nuclear Energy Society, London, 1984, Vol. 1, pp. 249–252.
60. M. G. Adamson, USAEC Report GEAP-14093, 1976.
61. G. Periaswami, T. Gnanasekaran, V. Ganesan, and C. K. Mathews, in: *Liquid Metal Engineering and Technology*, British Nuclear Energy Society, London, 1984, Vol. 1, pp. 399–404.
62. N. P. Bhat, K. Swaminathan, D. Krishnamurthy, O. M. Sreedharan, and M. Sundaresan, in: *Liquid Metal Engineering and Technology*, British Nuclear Energy Society, London, 1984, Vol. 1, pp. 323–328.
63. O. M. Sreedharan, B. S. Madan, and J. B. Gnanamoorthy, *J. Nucl. Mater. 119*, 296–300 (1983).
64. B. J. Shaiu, P. C. S. Wu, and P. Chiotti, *J. Nucl. Mater. 67*, 13–23 (1977).
65. C. F. Knights and B. A. Phillips, in: *High Temperature Chemistry of Inorganic and Ceramic Materials*, Spec. Publ. No. 30, The Chemical Society, London, 1977, pp. 134–145.
66. W. Dai, S. Seetharaman, and L. I. Staffansson, *Met. Trans. 15B*, 319–324 (1984).
67. P. G. Gadd and H. U. Borgstedt, in: *Liquid Metal Engineering and Technology*, British Nuclear Energy Society, London, 1984, Vol. 2, pp. 107–112.
68. M. G. Barker and G. A. Fairhall, *J. Chem. Res. 3*, 371 (1979).
69. M. Janson and R. Hoppe, *Z. Anorg. Allg. Chem. 408*, 104–106 (1974).
70. S. A. Frankham, Materials Behaviour in the Liquid Alkali Metals Lithium and Sodium, Ph.D. thesis, University of Nottingham, 1982.
71. A. J. Hooper and E. A. Trevillion, *J. Nucl. Mater. 48*, 216–222 (1973).
72. M. G. Barker, *Rev. Int. des hautes temp. et de refractaires 16*, 237–243 (1979).
73. M. G. Barker, G. A. Fairhall, and S. A. Frankham, in: *2nd Intern. Conf. on Liquid Metal*

Technology in Energy Production (J. M. Dahlke, Ed.), National Techn. Information Service, Springfield, Va., 1980, (CONF-800401 P2), Papers 18–41.
74. Y. Ikeda, H. Ito, T. Mizuno, and G. Matsumoto, *J. Nucl. Mater. 105*, 103–112 (1982).
75. T. Gnanasekaran, private communication, 1985.
76. J. M. Reau, C. Fouassier, and P. Hagenmuller, *Bull. Soc. Chim. France*, No. 11, 3827–3829 (1970).
77. H. Kessler, A. Hatterer, and C. Ringenbach, in: *Liquid Alkali Metals*, The Chemical Society, London, 1970, pp. 465–473.
78. J. P. Parant, R. Olagwaga, M. Devalette, C. Fouassier, and P. Hagenmuller, *J. Solid State Chem. 3*, 1–11 (1971).
79. C. Fouassier, C. Delmas, and P. Hagenmuller, *Mater. Res. Bull. 10*, 443–450 (1975).
80. M. G. Barker, S. A. Frankham, P. G. Gadd, and D. A. Moore, in: *Material Behavior and Physical Chemistry in Liquid Metal Systems* (H. U. Borgstedt, Ed.), Plenum Press, New York, 1982, pp. 113–120.
81. J. Jung, A. Reck, and R. Ziegler, *J. Nucl. Mater. 119*, 339–350 (1983).
82. J. K. Fink, in: *Reviews on Coatings and Corrosion 3*, 1–47 (1978).
83. M. G. Barker, P. G. Gadd, and M. J. Begley, *J. Chem. Soc., Chem. Comm.*, 379–381 (1981).
84. M. G. Barker, P. G. Gadd, and S. C. Wallwork, *J. Chem. Soc., Chem. Comm.*, 516–517 (1982).
85. J. Konys and H. U. Borgstedt, *J. Nucl. Mater. 131*, 158–161 (1985).
86. M. G. Barker and C. W. Morris, *J. Less-Common. Metals 44*, 169–176 (1976).
87. C. B. Alcock, K. T. Jacob, and S. Zador, *At. Energy Rev. Spec. Issue 6*, 7–65 (1976).
88. R. P. Beyer, K. C. Bennington, and R. R. Brown, *J. Chem. Thermodynamics 17*, 11–17 (1985).
89. J. Hauck, *J. Inorg. Nucl. Chem. 36*, 2291–2298 (1974).
90. S. Kemmler-Sack and W. Rüdorff, *Z. Anorg. Allg. Chem. 354*, 255–272 (1967).
91. D. W. Osborne, H. K. Flotow, and H. R. Hoekstra, *J. Chem. Thermodynamics 6*, 751–756 (1974).
92. M. G. Adamson, E. A. Aitken, and D. W. Jeter, in: *Intern. Conf. on Liquid Metal Technology in Energy Production* (M. H. Cooper, Ed.), National Techn. Information Service, Springfield, Va., 1976, (CONF-760503), Vol. 2, pp. 866–874.
93. M. A. Mignanelli and P. E. Potter, *J. Nucl. Mater. 114*, 168–180 (1983).
94. M. A. Mignanelli and P. E. Potter, *J. Nucl. Mater. 125*, 182–201 (1984).
95. M. G. Adamson, M. A. Mignanelli, P. E. Potter, and M. H. Rand, *J. Nucl. Mater. 97*, 203–212 (1981).
96. M. A. Mignanelli and P. E. Potter, *J. Nucl. Mater. 130*, 289–297 (1985).
97. M. Housseau, G. Dean, J. P. Marcon, and J. F. Martin, Report CEA-N-1588, CEA, Paris, 1973.
98. A. B. Van Egmond and E. H. P. Cordfunke, *J. Inorg. Nucl. Chem. 38*, 2245–2247 (1976).
99. E. H. P. Cordfunke, A. B. Van Egmond, and G. Van Voorst, *J. Inorg. Nucl. Chem. 37*, 1433–1436 (1975).
100. A. B. Van Egmond, *J. Inorg. Nucl. Chem. 38*, 1645–1647, 1649–1651, 2105–2107 (1976).
101. D. C. Fee and C. E. Johnson, *J. Nucl. Mater. 99*, 107–116 (1981).
102. R. Hoppe, *Bull. Soc. Chim. France 4*, 1115–1121 (1965).
103. R. Clos, M. Devalette, P. Hagenmuller, R. Hoppe, and E. Palletta, *C. R. Acad. Sci., Ser. C 265*, 801–804 (1977).
104. G. Blasse, *J. Inorg. Nucl. Chem. 28*, 2444–2445 (1966).
105. R. Clos, M. Devalette, C. Fouassier, and P. Hagenmuller, *Mater. Res. Bull. 5*, 179–184 (1970).

106. M. A. Mignanelli and P. E. Potter, *J. Nucl. Mater. 97*, 213–222 (1981).
107. M. G. Barker, S. A. Frankham, and P. G. Gadd, *J. Inorg. Nucl. Chem. 43*, 2815–2819 (1981).
108. M. V. Krishnaiah and P. Sriramamurti, *J. Am. Ceran. Soc. 67*, 568–571 (1984).
109. T. L. Markin and E. J. McIver, *Plutonium 1965* (A. E. Kay and M. B. Waldron, Eds.), Chapman and Hall, London, 1967.
110. R. E. Woodley and M. G. Adamson, *J. Nucl. Mater. 82*, 65–75 (1979).
111. R. E. Woodley, *J. Am. Ceran. Soc. 56*, 116–119 (1973).

6

Carbon and Nitrogen in Alkali Metals

Carbon and nitrogen dissolved in liquid alkali metals are very important in the corrosion of steels and refractory alloys by liquid metals. Both elements are of importance even as minor alloying elements. They influence the mechanical properties of the materials, mainly their high-temperature strength and ductility. Alkali metals containing dissolved carbon or nitrogen tend to exchange these elements with solid materials. Thus, ferritic steels may be decarburized or denitrided at high temperatures, while vanadium alloys might be carburized or nitrided when exposed to the same alkali metal melt. The knowledge of the driving forces for exchange of the interstitial elements is necessary for the understanding of corrosion and mass transfer phenomena.

6.1. Carbon Compounds of Alkali Metals

Alkali metals form compounds with carbon, which may be of the acetylide type M_2C_2[1] or lamellar graphite compounds C_xM, in which several relationships of carbon and alkali atoms can be observed.[2] The acetylides of the light alkali metals are more or less stable compounds which are formed in mixtures containing high amounts of the metallic elements. Therefore, it is assumed that the acetylides are those compounds which are in equilibrium with carbon dissolved in the alkali metal melts. The formation of lamellar species is often observed in high-temperature reactions of graphite with heavier alkali metals.

The alkali metal acetylides Li_2C_2, Na_2C_2, and K_2C_2 have been prepared by several methods. The chemical stability of these ionic compounds diminishes with increasing atomic number of the alkali atom. Lithium acetylide is formed by the reaction of lithium and carbon in an excess of the liquid metal according to equation (1)

$$2Li + 2C \xrightleftharpoons{\text{liq. Li}} Li_2C_2 \tag{1}$$

It crystallizes in the primitive monoclinic lattice with 19 molecules of Li_2C_2 per unit cell.[3] The lattice parameters of Li_2C_2 are $a = 7.801$ Å; $b = 8.815$ Å; $c = 10.865$ Å; and $\beta = 76.8 \pm 0.1°$. Measurements of the carbon activity in solutions of lithium and lithium acetylide in liquid sodium[4] indicate that the acetylide dissociates into its constituents in such solutions. Measurements of the electromotive forces in the same solutions allow an estimation of the free energy of formation of the lithium acetylide at 600°C. The value of $\Delta G_{873} = -89$ kJ/mol indicates the chemical stability of this compound compared to other alkali acetylides.

The structures of the sodium and potassium acetylides, white crystalline compounds which can be obtained by thermal decomposition of the MHC_2 compounds *in vacuo*, are similar to the tetragonal structure of the peroxide K_2O_2, as seen from their X-ray powder diffraction patterns.[5] Sodium acetylide is less stable than its lithium analogue. Its free energy of formation has been determined to be $\Delta G_{298} = -21.0$ kJ/mol. Potassium acetylide should be still less stable. Its free energy of formation has never been published, and does not appear to have been determined so far. Sodium acetylide might be stabilized by dissolution in the liquid metals since the free energy of solution is very high. The heavy alkali metals rubidium and cesium form acetylides which are still less stable.

The acetylides of lithium and sodium evolve acetylene when their solutions in the alkali metals are hydrolyzed by reaction with water. The absorption of acetylene in liquid lithium results in the formation of lithium acetylide and hydride[6]:

$$4Li + C_2H_2 \rightleftarrows Li_2C_2 + 2LiH \tag{2}$$

The structure of the interlamellar compounds in the graphite–alkali metal systems has been studied in detail by Rüdorff.[2] The reactions of the three heavier alkali metals with powdered graphite at temperatures above ~400°C result in the penetration of the alkali metals which enter the planes between the crystallographic planes, thus expanding the graphite lattice. The interlamellar compounds are able to take up considerable amounts of alkali atoms into their structures. The existence of stoichiometric compounds with carbon/alkali metal atomic ratios of 8, 16,

24, 36, 48, and 60 has been shown for compounds of potassium, rubidium, and cesium.

A layer of alkali metal atoms is present between each pair of graphite planes in the compounds of the C_8M type. In compounds which contain smaller amounts of alkali metals there are two graphite planes between the alkali metal layers. The formation of the interlamellar compounds causes an increase in the interplanar distances to 5.41 Å for potassium, 5.61 Å for rubidium, and5.95 Å for cesium. In the C_8M compound the alkali metal atoms form a triangular net with a distance of $a = 4.91$ Å between the alkali atoms. The crystal lattice of potassium graphite C_8K is shown in Fig. 6.1.

The compound of the composition $C_{24}M$ has only half the number of alkali metal atom layers; as well, the number of alkali atoms in each layer is reduced by one-third. This fact causes the formation of a hexagonal network of alkali atoms in these layers with the same distances between the atoms as in the C_8M-type compounds. For potassium graphite the lattice constant c increases with decreasing alkali metal content from $c = 5.41$ Å for C_8K to $c = 8.76$ Å for $C_{24}K$ to $c = 18.83$ Å for the compound $C_{60}K$.

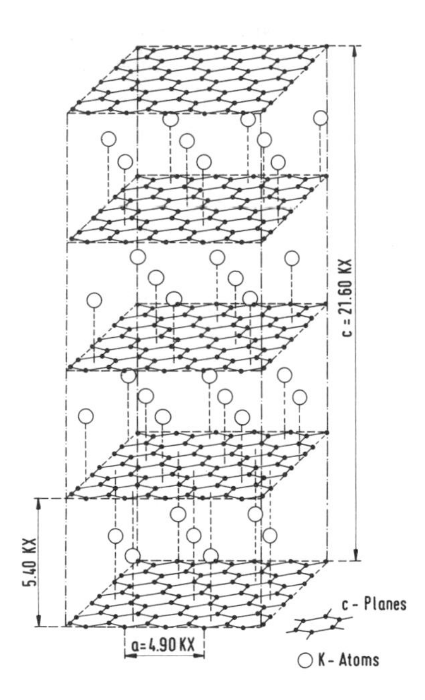

Figure 6.1. The crystal lattice of potassium graphite C_8K (after ref. 2; reproduced with the permission of Verlag Johann Ambrosius Barth, Leipzig, GDR).

Potassium graphite C_8K has a heat of formation of $\Delta H = -32.1$ kJ/mol. The energy of the reaction

$$C_{24}K + 2K \rightleftarrows 3C_8K \qquad (3)$$

has been determined to be $\Delta H = -28.0$ kJ/mol. The ΔH values for the rubidium and cesium compounds indicate that these graphite compounds are still more stable: $\Delta H = -47.8$ kJ/mol for C_8Rb, and $\Delta H = -83.2$ kJ/mol for C_8Cs.

Alkali metal–graphite compounds are highly reactive. They ignite in contact with air and react explosively with water. When hydrolyzed with water under a protective atmosphere, the formation of hydrogen gas and alkali hydroxide is observed; acetylene or other hydrocarbons do not form. The compound C_8K possesses a weak paramagnetism, which corresponds to the paramagnetism of the electron gas in a metal and does not depend on the temperature. The potassium–graphite compound has a considerably higher electrical conductivity than graphite. The temperature coefficient of conductivity in the compound is as negative as in metals. The alkali metals add their free electrons to the conduction band of graphite, thus causing the higher metallic conductivity.

Experiments indicate that sodium is also able to form such compounds with the approximate formula $C_{64}Na$. The graphite compound which has the highest content of alkali metal (C_8M) decomposes under vacuum at higher temperatures to yield alkali metal vapor and compounds with lower alkali metal content:

$$3C_8K \rightleftarrows C_{24}K + 2K \qquad (4)$$

The graphite compounds of the metals rubidium and cesium play a role in alkali metal technology. Their formation offers a way to absorb these fission elements out of the primary sodium of a fast sodium-cooled reactor. It has been shown that graphite immersed in liquid sodium has the ability to trap cesium and rubidium in the coolant circuit of the reactor.[7]

It has been shown in a recent study[8] that graphite has also the capacity to form intercalation compounds with lithium in a solid–liquid reaction. The degassed and highly oriented pyrolytic graphite in the size of a cube is immersed in a pool of molten lithium at 350°C, where the compound LiC_6 is formed. A compound of the composition LiC_{18} is achieved by immersion of the graphite cube in a liquid Na–3.5% Li alloy. At a temperature above 400°C the formation of lithium acetylide is favored. The kinetics of the intercalation are rapid. After the reaction the surfaces of the cubes are cleaned. They have to be taken out of the pools under protective atmosphere. The compound LiC_6 has a characteristic golden yellow color, while the compound LiC_{18} is steel blue. X-ray diffraction shows that the spacing of the hexagonal planes is 3.706 Å for LiC_6 and 7.044 Å for LiC_{18}.

6.2. Alkali Metal Nitrides

In contact with liquid lithium, nitrogen forms the nitride Li_3N.[9] This compound can be considered as having a typical ionic structure, N^{3-} being coordinated by eight Li^+ ions in a regular way. It has a hexagonal crystal structure with $a = 3.648$ Å, $c = 3.875$ Å, $Z = 1$, and the space group has been determined to be *P6/mmn*.[10] The phase diagram of the lithium–lithium nitride system is shown in Fig. 6.2. This salt, which melts at a temperature of 815°C, has remarkable thermodynamic stability as is seen from its free energy of formation, $\Delta G_{298} = -41.2$ kJ/mol.

There are no compounds known in the sodium–nitrogen system. The very low amounts of nitrogen soluble in liquid sodium seem to be present as the dissolved element. The nitride Na_3N may be formed—as a metastable compound—only by means of electrical discharges in sodium–nitrogen mixtures. Sodium nitride tends to react with excess nitrogen, and the product of this secondary reaction is sodium azide, NaN_3. The formation of nitrides of potassium, rubidium, and cesium has not been reported. Their existence is unlikely.

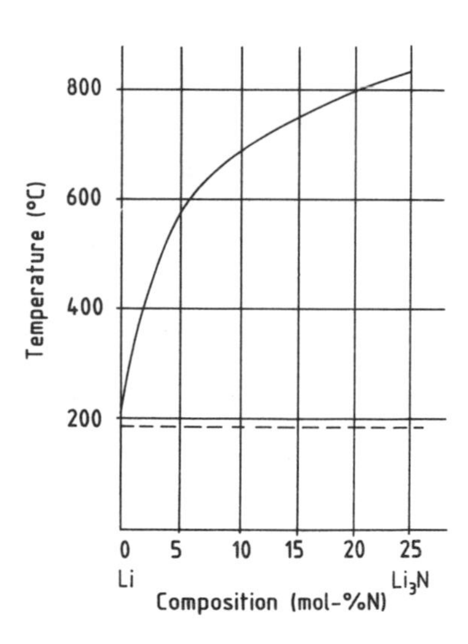

Figure 6.2. The lithium–lithium nitride phase diagram (after ref. 11; reproduced with the permission of Elsevier Sequoia, Oxford, UK).

6.3. Compounds of Alkali Metals with Carbon and Nitrogen

Carbon dissolved in liquid lithium reacts with nitrogen or with lithium nitride at a temperature of ~475°C:

$$4Li_3N + Li_2C_2 \rightleftarrows 2Li_2NCN + 10Li \tag{5}$$

The reaction product, dilithium cyanamide, is chemically stable and can be isolated from the reaction mixture by vacuum distillation of the excess metal. The free energy of formation of this crystalline compound has to be more negative than $\Delta G^0 = -359$ kJ/mol.

Reactions of carbon and nitrogen in liquid sodium are different. The elements need a temperature of about 700°C to react in this molten metal. The reaction product is sodium cyanide, NaCN.(12) Sodium cyanide is stable in contact with sodium metal and remains as the residue of vacuum distillation up to a temperature of 600°C. Its stability is much higher than that of the unstable nitride, Na_3N, and the metastable acetylide, Na_2C_2. This fact favors its formation in the presence of molten sodium. It is assumed that sodium cyanide is also dissolved by liquid sodium, since high nitrogen pressure is found to raise carbon solubility in the metal. Sodium cyanide is precipitated in the cold traps of sodium loops, where it is detected among other carbon-bearing compounds. This indicates its formation under loop conditions.

The chemical stability of potassium cyanide allows the conclusion that the same compound formation as in sodium occurs in the potassium–carbon–nitrogen system. Compounds of the two heaviest alkali metals with both elements in contact with an excess of the metals have not been reported.

6.4. Solubility of Carbon in Alkali Metals

Only two studies on the solubility of carbon or acetylide in liquid lithium have been published so far.(13,14) The more recent publication is based on the estimation of carbon by means of the acetylene evolution method.(14) A regression curve has been calculated from the results of these analyses, giving the saturation solubility equation:

$$\log[C]_{s(Li)}(\text{wppm}) = 5.9894 - \frac{2645.5}{T(K)} \tag{6}$$

The earlier work,[13] based on X-ray measurements, has generated considerably higher values for carbon solubility in liquid lithium. Analytical problems and impurity interferences are the most probable reasons for the discrepancies. Some additional measurements of acetylide solubility in very pure lithium with varying nitrogen content would be very useful. Cold-trap studies made in a lithium loop[15] support the saturation solubility equation given by equation (6).

The same authors have recently made a study of the solubility of carbon in lithium with a nitrogen content below 50 wppm.[16] They used the same apparatus and analytical procedure as in the previous work. There is a remarkable effect of nitrogen on the solubility of carbon in lithium. In the very pure liquid metal the solubility of carbon is in accord with equation (6a).

$$\log[\mathrm{C}]_{s(\mathrm{Li})}(\mathrm{wppm}) = 7.459 - \frac{3740}{T(\mathrm{K})} \tag{6a}$$

A lot of work has been devoted to the solubility of carbon in liquid sodium. The early studies did not give a clear picture of the saturation concentrations of carbon or acetylide in the metal as a function of temperature. However, the more recent and extensive studies published by Ainsley *et al.*[17] and by Longson and Thorley[18] have given results that agree well with each other, though they differ from the results of most of the earlier investigations. The equations given in these publications are:

$$\log[\mathrm{C}]_{s(\mathrm{Na})}(\mathrm{wppm}) = 7.646 - \frac{5970}{T(\mathrm{K})} \tag{7a}$$

and

$$\log[\mathrm{C}]_{s(\mathrm{Na})}(\mathrm{wppm}) = 7.20 - \frac{5465}{T(\mathrm{K})} \tag{7b}$$

These solubility data are in agreement with carbon meter measurements in sodium loops as well as with carbon exchange phenomena observed in stainless steel–sodium systems.

The solubilities of carbon in molten potassium and rubidium have not yet been determined. The solubility of carbon in cesium has been estimated at two temperatures. The data seem to be insufficient and have not yet been critically checked. Therefore, equation (8), calculated from these data, has to be regarded as tentative.[19]

$$\log[\mathrm{C}]_{s(\mathrm{Cs})}(\mathrm{wppm}) = 2.8632 - \frac{1050}{T(\mathrm{K})} \tag{8}$$

The temperature gradient of this solubility curve differs considerably from that for the systems in which the light alkali metals are involved. The three saturation curves of carbon in the alkali metals lithium, sodium, and cesium are compared in Fig. 6.3.

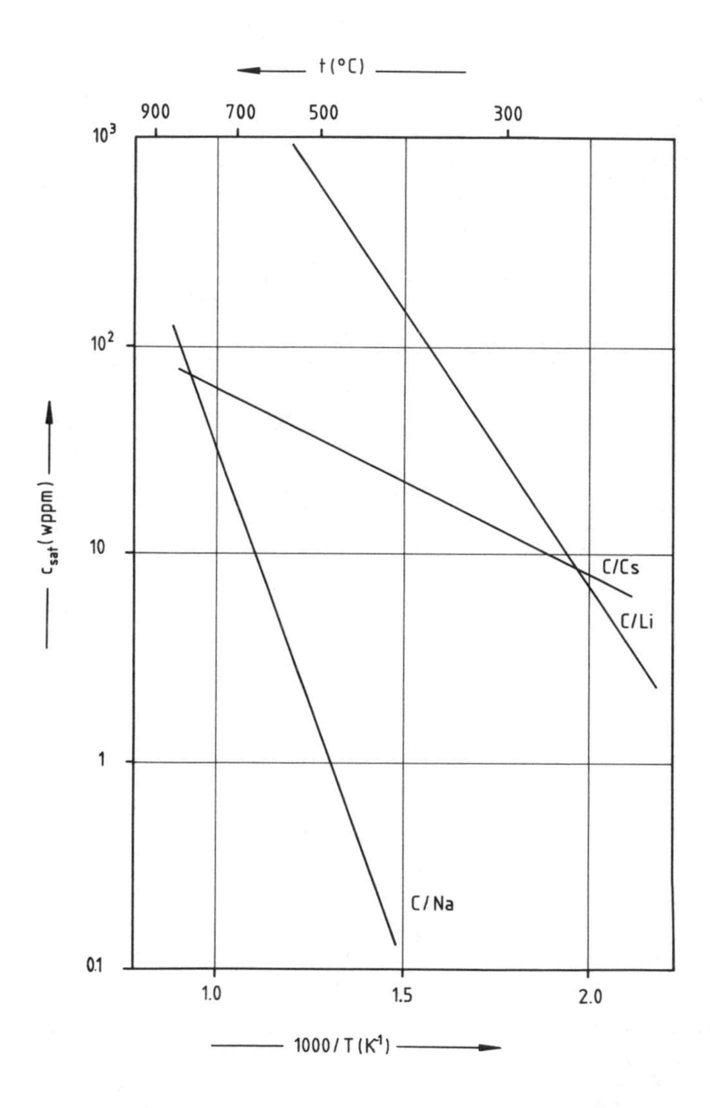

Figure 6.3. Comparison of the saturation concentrations of carbon in lithium, sodium, and cesium.

6.5. Solubility of Nitrogen in Alkali Metals

The only stable binary alkali nitrogen compound, lithium nitride, Li_3N, is fairly soluble in liquid lithium, as can be concluded from the phase diagram. Two groups applying different methods have established solubility curves. Yonco *et al.*[20] have determined nitrogen in saturated solutions by applying the micro Kjeldahl method, while Adams *et al.*[11] used measurements of the electrical resistivities of lithium absorbing nitrogen from the gas phase. The saturation equations derived from their data are in excellent agreement.

$$\log[\mathrm{N}]_{s(\mathrm{Li})}(\mathrm{wppm}) = 7.597 - \frac{2098.5}{T(\mathrm{K})} \qquad \text{(Ref. 20)} \qquad (9a)$$

$$\log[\mathrm{N}]_{s(\mathrm{Li})}(\mathrm{wppm}) = 7.521 - \frac{2047.3}{T(\mathrm{K})} \qquad \text{(Ref. 11)} \qquad (9b)$$

The two solubility plots are compared in Fig. 6.4.

The solubility of nitrogen in liquid sodium is extremely low, comparable to the solubilities of noble gases in this metal. Therefore, and because of the instability of binary sodium nitrogen compounds, it is concluded that nitrogen dissolves in the liquid metal in its molecular state. The solubility of the element can be expressed by

$$\log[\mathrm{N}]_{s(\mathrm{Na})}(\mathrm{wppm}) = -7.17 - \frac{2780}{T(\mathrm{K})} \qquad (10)$$

as determined by Veleckis *et al.*[21] Measurements of the solubility of nitrogen in the three other alkali metals have not been published so far. The solubility in potassium, rubidium, and cesium has to be expected to be of the same order as in sodium.

6.6. Analysis of Carbon in Alkali Metals

Analytical methods to determine carbon concentrations in alkali metals include the application of monitors as different types of carbon meters and of tabs which have been equilibrated with alkali metals with respect to carbon. Another method used in laboratories is the analysis of the residues of alkali metal vacuum distillations.

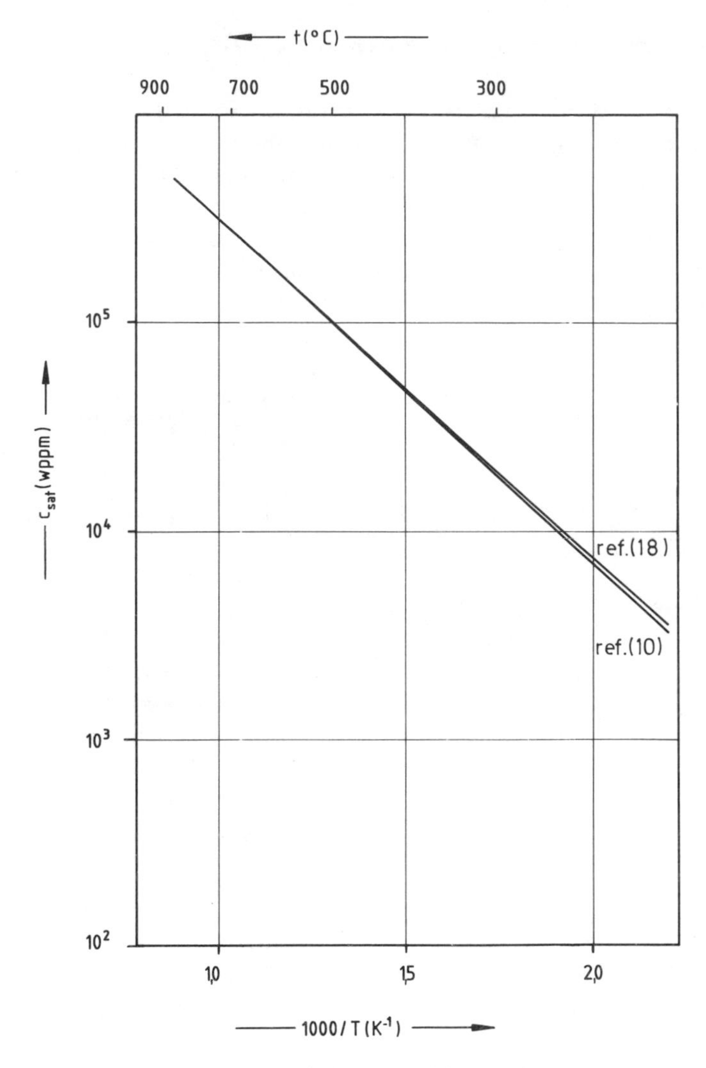

Figure 6.4. Solubility curves of nitrogen in lithium (after refs. 11 and 20).

6.6.1. The Distillation Method

Alkali metal samples containing higher amounts of carbon have been used to determine carbon after a distillation procedure. The analytical distillation serves to separate the alkali metal solvent from carbon or carbon compounds in order to apply combustion analysis to the residue. Crucibles of heat-resistant ceramic materials like alumina are applicable for sample collection as well as for distillation and direct use in the combustion apparatus, the central part of which is a high-temperature furnace. The ceramic crucibles have to be heat-treated before their use in this technical procedure. They have to be surrounded by metallic clads during the vacuum distillation. These clads are necessary to improve the electrical coupling of the induction coils to the content of the crucibles. The sampling procedure is essentially the same as for the analysis of oxygen in alkali metals.

The method can be applied for the estimation of carbon in sodium and the heavier alkali metals. Lithium, however, reacts with alumina crucibles at the distillation temperature. Armoc iron with low carbon content can be recommended as a material from which distillation vessels for use with lithium can be made. The sensitivity of the method allows the detection of 2 wppm of carbon in samples with a weight of about five grams. The distillation method has the advantage that not only elemental or acetylide carbon, but also carbon in compounds such as carbonate, cyanide, and even carbides may be estimated.[22] The sensitivity, however, is not sufficient to determine carbon in dilute solutions in the alkali metals.

6.6.2. Analysis by Means of Hydrolysis of Alkali Metals

Alkali metals have to be dissolved in water which is kept at low temperature by means of intensive cooling and under inert atmosphere. The product of this procedure is a solution of alkali hydroxide in water. Lithium acetylide decomposes with evolution of acetylene gas during this hydrolysis. The evolved gas may be measured using a gas chromatograph. The analysis is highly sensitive. The presence of larger amounts of nitrogen, however, may disturb the evolution of acetylene since cyanamide may be formed in a side reaction.

Solutions of carbon in sodium do not evolve quantitatively acetylene gas; hydroxide solutions may contain some carbon as dissolved sodium cyanide. This part has to be evolved by means of a reaction with pure concentrated sulfuric acid and subsequent carrier gas distillation. The distilled HCN is absorbed in an alkaline solution, in which it may be estimated through a potentiometric titration using an ion-sensitive electrode. A glass apparatus for hydrolysis and distillation is recommended for the several steps of this analytical procedure.[23] The sodium samples (or samples of heavier alkali metals) are collected in crucibles of sintered alumina which have been purified by annealing at 1000°C for several hours. The method is more sensitive than the vacuum distillation technique.

6.6.3. Foil Equilibration Methods

The chemical exchange of carbon between alkali metals and solid metals or alloys as stainless steels or binary iron alloys is fast enough to attain equilibria in not too long reaction times at sufficiently high temperatures (~700°C). The solid materials have to be inserted into the liquid metals as thin foils (or even wires) in order to keep the diffusion paths (and the reaction times) small.

In the sodium–austenitic steel system, experience has demonstrated that one can rely upon the results gained under conditions decarburizing the steels.[24] The materials of the tabs, in which the relationships of carbon concentrations to chemical activities of carbon are already established, are alloys such as the austenitic steel Fe–18Cr–9Ni and the binary alloys Fe–8Ni and Fe–20Mn.[25]

The equilibration takes place at temperatures from 550 to 750°C, when tabs are inserted into the alkali metals for exposure times between 100 and 1000 hours. The foils have to be cleaned carefully after removal from the alkali metals. The tabs are analyzed in the conventional way. The results of the carbon determinations are used for the calculation of the chemical activities of carbon in the equilibrated foils, which should be equal to the activity of carbon in the alkali metal melts. The activites corresponding to several concentrations of carbon in the steel Fe–18Cr–9Ni and in molten sodium are demonstrated in Fig. 6.5. The sensitivity of this analytical method depends on the tab material and its mass. The high carbon content at equilibrium in the austenitic steels is the reason

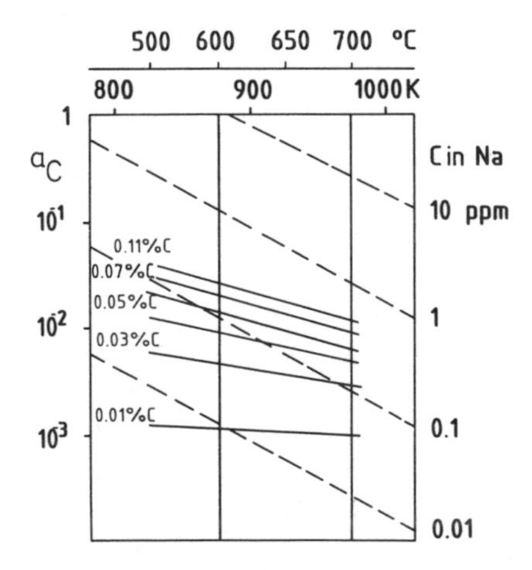

Figure 6.5. Equilibrium concentrations of carbon in the austenitic steel Fe–18Cr–9Ni and in liquid sodium at temperatures between 800 and 1000 K.

for the high sensitivity of the estimation of carbon in alkali metals if these tab materials are used. Thus, even carbon concentrations which are much lower than the detection limit of the distillation analysis are within the sensitivity range of the foil equilibration method. The advantage of this method is that its application avoids the danger of contamination of alkali metal samples. However, the foils have to be protected against the atmosphere after the exposure to the alkali metals.

The foil equilibration method is well established for carbon determinations in liquid sodium, where the results of this technique are in fair agreement with results of monitor measurements. The foil equilibration has also been applied in analyses of liquid lithium, and its application in analyses of molten heavy alkali metals seems to be possible.

6.6.4. The Diffusion Carbon Meter

The principle of the diffusion carbon meter, the most successful version of which is the Harwell carbon meter,[(26)] is based on the diffusion of carbon from the alkali metal through a thin metallic membrane into a gas system, where it is estimated by means of a conventional gas analysis. A long iron capillary immersed in the molten metal serves as the membrane. The inner surface of the capillary is covered by a layer of iron oxide, FeO, which reacts with the carbon passing through the iron wall. Carbon monoxide is the product of this reaction. A flow of inert gas (argon) passing through the capillary sweeps the carbon monoxide into the gas analysis unit. At the entrance of this device, CO is converted into methane which is continuously measured by a pair of flame ionization detectors.

The formation of carbon monoxide keeps the carbon activity on the inner surface of the iron membrane at a low level. Thus, a steady carbon flow is created, which is directed from the sodium through the membrane to the gas analytical unit. This carbon flux is proportional to the carbon potential in the alkali metal. A calibrated leak enables the meter to relate the measured carbon flux to an internal standard. Figure 6.6 shows the principle of the carbon meter probe and the schematic gas flow in the meter. In sodium loops with cold-trap operation temperatures of $\sim 125°C$, the meter has the sensitivity to estimate carbon activities as low as 2.0×10^{-3} at 550°C corresponding to carbon concentrations in the 10-ppb level.

The Harwell carbon meter can be applied in all the liquid alkali metals, though only the experience gained with liquid sodium has been reported so far. Its application depends on the compatibility of the membrane material with the molten metals. Lithium containing nitrogen attacks

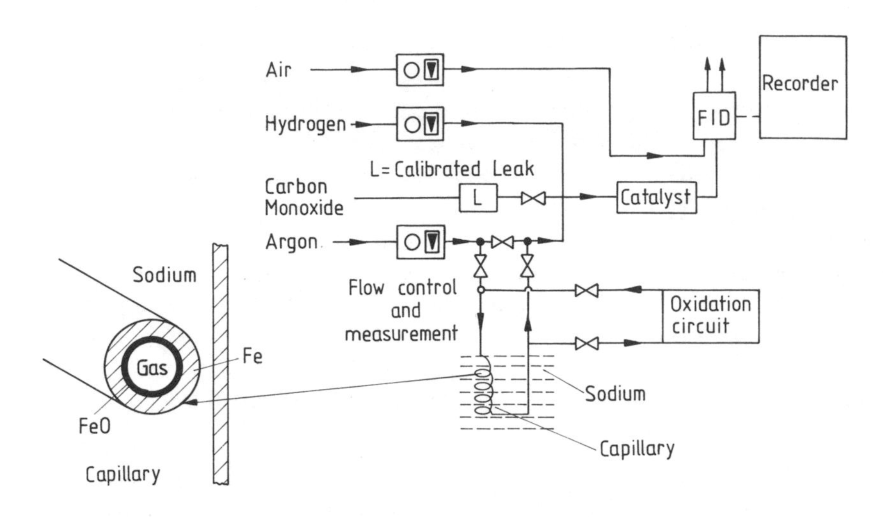

Figure 6.6. Principle and schematic flow diagram of the Harwell carbon meter.

the iron capillaries severely. Iron, however, is compatible with very pure lithium as well as with the heavier alkali metals. It can be an instrument to analyze the carbon chemistry in molten potassium, rubidium, and cesium.

6.6.5. The Electrochemical Carbon Meter

Electrochemical carbon meters are extremely useful for the estimation of absolute chemical activities of carbon in liquid alkali metals. They are based on electrochemical concentration cells or, more precisely, activity cells. The carbon potential difference between the liquid metal and a reference electrode generates a voltage which can be used to estimate the carbon potential in the alkali metal, if the potential at the reference electrode is known.

The electrolyte is a molten salt mixture which has sufficiently high conductivity for carbon ions. The most promising electrochemical meter uses the eutectic mixture of lithium carbonate with sodium carbonate,[27] which melts at approximately 500°C, as electrolyte. The molten salt mixture, which has to be free from humidity or other impurities, is contained in a thin-walled iron membrane cup. The reference electrode is graphite or any other material with a well-defined carbon potential. It is immersed in the molten salt electrolyte. The iron cup is placed into the liquid metal (Fig. 6.7). The iron equilibrates with the liquid metal with

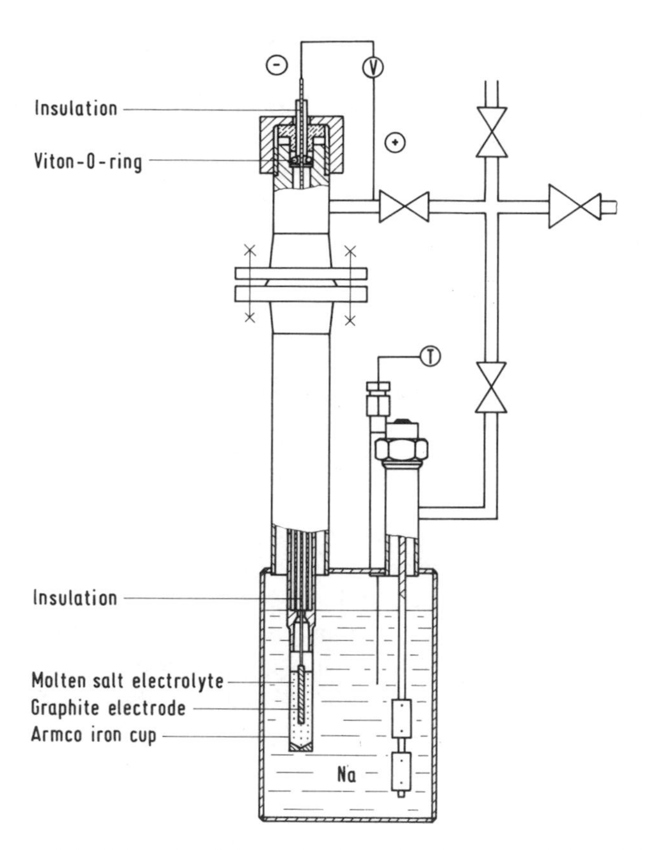

Figure 6.7. Schematic of the electrochemical carbon meter (after ref. 28).

respect to carbon. The Nernst equation gives the potential of the electrochemical chain:

$$E = \frac{RT}{nF} \cdot \ln \frac{a_C^M}{a_C^{ref}} \tag{11}$$

where R is the gas constant, T the temperature in K, F the Faraday constant, and $n = 4$, the number of electrons involved.

Electrochemical carbon meters have not yet been developed to a state in which they can be applied for long-time measurements in loops. They have, however, been successfully tested in laboratory-scale setups.[29]

A combination of the principles of the diffusion-type carbon meter with an electrochemical cell is based on the electrochemical measurement of the oxygen activity of a carbon monoxide/carbon dioxide gas mixture.[30] The oxygen potential in the gas is estimated by means of a solid-electrolyte oxygen meter applying zirconia–calcia electrolyte tubes and tin–tin oxide reference electrodes. The gas volume around the oxygen meter consists of 66 parts CO and 34 parts CO_2 at 2.5 mbar pressure. The whole setup is placed in an iron cylinder with a diffusion membrane as wall. This cylinder is immersed in the alkali metal. Carbon diffusing through the iron membrane into the gas reduces the oxygen potential by means of the reaction

$$C_{(Fe)} + CO_2 \rightleftarrows 2CO \tag{12}$$

The oxygen activity of the gas mixture in the carbon meter is established by the equilibrium

$$CO_2 \rightleftarrows CO + \tfrac{1}{2}O_2 \tag{13}$$

The oxygen potential in the gas volume is thus determined by the ratio of CO_2 to CO, and this ratio depends on the amount of carbon in the iron membrane, which is in contact with both the alkali metal and the gas. Thus, the meter should be capable of measuring the absolute carbon potential values in the alkali metals. Its sensitivity seems to be very high, and the use of this meter should not be limited to liquid sodium, for which it has been developed.

6.7. Analysis of Nitrogen in Alkali Metals

Methods to estimate nitrogen concentrations are mainly developed for lithium, in which the amounts of dissolved nitrogen are in the percent range. The determination of the nonmetal in liquid sodium has also been reported, though concentrations are extremely low in the sodium–nitrogen system.

6.7.1. Determination of Nitrogen by Means of the Evolution of Ammonia

A titrimetric method is applicable to determine nitrogen dissolved in lithium as nitride. The method is based on the evolution of ammonia, resulting from hydrolysis of the nitride, when the liquid metal is dissolved in water. The ammonia is separated from the hydrolyzed sample by means of distillation into a second flask, where it is dissolved in a dilute solution of boric acid. This solution is titrated with 0.00429 *N* hydrochloric acid to the neutral end point after the addition of an indicator.[31] The same method has been applied to determine traces of nitrogen dissolved in sodium. The procedure has to be performed under a protective atmosphere of dry and pure argon; in addition to nitrogen, other atmospheric elements also have to be excluded.

The evolved ammonia is collected by absorption in a dilute ultra-pure sulfuric acid in another version of the Kjeldahl method. In this solution, ammonia reacts with sodium phenolate to form indophenol, and the blue color of this solution is used for the photometric determination.[32] The setup is similar to the ordinary Kjeldahl apparatus generally used for nitrogen determinations. The method is very accurate and sensitive. However, the hydrolysis only evolves the nitrogen present as nitride. Nitrogen in compounds like cyanide or cyanamide does not react to form ammonia when the alkali metals are dissolved in water.

6.7.2. The Foil Equilibration Method

Nitrogen exchange between stainless steel and alkali metals leads to chemical equilibration of nitrogen between the two phases. This is applied to estimate nitrogen in the alkali metals. The basis of this method is the established relationship between the chemical activity and the analytical concentration of nitrogen in an 18-9 chromium–nickel steel at temperatures between 550 and 700°C. It has been demonstrated that the equilibration of stainless steel foils with liquid sodium and lithium with respect to nitrogen generates values which are in fair agreement with other methods.[33] The foils of austenitic steel of 0.025- to 0.100-mm thickness have to be inserted into the liquid metal for 250 to 1000 hours at 700°C in order to reach chemical equilibrium and an equal distribution of nitrogen in the cross sections of the foils. The chemical activities of nitrogen in its dilute solutions in the steel have to be calculated under the assumption that the free energy of formation of chromium nitride, Cr_2N, has to be considered, since this is the stable compound in the steel and precipitates in supersaturated solutions. The very low solubility of nitrogen in liquid sodium is the reason for the considerably high chemical activities of nitrogen even at extremely low concentrations, which cannot be detected by means of wet chemical analyses. Therefore, foil equilibration is the only method applicable to unsaturated solutions of nitrogen in sodium. The foil equilibration is one method among others for the analysis of nitrogen in lithium. Its advantage is that samples do not have to be drawn out of the container. Foil equilibration might be disturbed by the presence of carbon in lithium, which causes the formation of cyanamide. Thus, the activity of nitrogen in lithium is lowered.

The results of foil equilibration techniques are easier to understand and interpret if the foils are denitrided. The precipitation of nitrides in the steels due to the nitriding reaction of alkali metals causes difficulties in the estimation of chemical activities. Therefore, the method should be applied to determinations of low contents of nitrogen in alkali metals, while in solutions with higher concentrations direct sampling and analyses might be preferable.

6.7.3. The Determination of Nitrogen Concentrations by Means of Conductivity Measurements

Nitrogen can easily be monitored in liquid lithium and other alkali metals in an on-line method by means of measurements of the electrical resistivity of a volume of the melt in a capillary of known dimensions. A Wheatstone bridge is applied to measure the electrical resistance of the solutions. Resistivities may be calculated from the measured values and the capillary dimensions.[34] The increase of the resistivity is directly related to increase in the concentration of dissolved nitride. At 400°C, for example, an increase of 1 mol % nitrogen causes a resistivity change of $7.0 \times 10^{-8}\ \Omega m$. The sensitivity of the method is in the range of 0.01 mol % nitrogen, since resistivity changes of $0.5 \times 10^{-10}\ \Omega m$ can be detected.

The resistivity of the molten metals is also influenced by the presence of other nonmetallic impurities. Therefore, the method is not specific. It is applicable in systems in which nitrogen is the main impurity. Concentrations close to the level in gettered lithium can be estimated by means of resistivity changes. A nitrogen meter for practical application in lithium loops based on this method has not yet been developed.

6.8. Chemical Potentials of Carbon and Nitrogen in Alkali Metals

The chemical potentials of the dissolved nonmetals carbon and nitrogen depend on their concentration in relation to the saturation values and on the thermodynamic stability of the compounds formed in solution. The chemical stability of these compounds is related to the free enthalpy of formation. The temperature dependencies of the Gibbs free energy of formation are given in a recent survey.[35] The equation valid for Li_2C_2 is

$$\Delta_f G^0 = -59{,}434.1 + 47.88 \cdot T \tag{14}$$

and for the metastable sodium acetylide Na_2C_2

$$\Delta_f G^0 = -44{,}286.8 + 52.61 \cdot T \qquad (15)$$

(in J/mol with T in K). The Gibbs free energy of formation of the lithium nitride Li_3N can be expressed by

$$\Delta_f G^0 = -16{,}6719 + 142.63 \cdot T \qquad (16)$$

However, a critical check of the thermochemical data seems to be still necessary. Enthalpy data for the acetylides or nitrides of the other alkali metals are not available due to the fact that these compounds are more or less unstable.

References

1. Ch. E. Messer, in: *The Alkali Metals*, Spec. Publ. No. 22, The Chemical Society, London, 1967, pp. 183–198.
2. W. Rüdorff, in: *Advances in Inorganic Chemistry and Radiochemistry* (H. J. Emeléus and A. G. Sharpe, Eds.), Vol. 1, Academic Press, New York, 1959, pp. 223–266.
3. D. R. Secrist and L. G. Wisnyi, *Acta Cryst. 15*, 1042–1043 (1962).
4. R. J. Pulham and P. Hubberstey, *J. Nucl. Mater. 115*, 239–250 (1983).
5. H. Föppl, *Angew. Chem. 70*, 401 (1958).
6. R. J. Pulham, P. Hubberstey, A. E. Thunder, A. Harper, and A. T. Dadd, in: *2nd Intern. Conf. on Liquid Metal Technology in Energy Production* (J. M. Dahlke, Ed.), National Techn. Information Service, Springfield, Va., 1980 (CONF-800401-P2), Vol. 2, 18-1.
7. J. C. Clifford, J. M. Williams, and J. C. McGuire, in: *Alkali Metal Coolants*, International Atomic Energy Agency, Vienna, 1967, pp. 759–779.
8. S. Basu, G. K. Wertheim, and S. B. Dicenzo, in: *Lithium—Current Applications in Science, Medicine, and Technology* (R. O. Bach, Ed.), John Wiley and Sons, New York, 1985, pp. 187–194.
9. K. A. Bolshakov, P. I. Fedorov and L. Stepina, *Izv. Vysh. Ucheb. Zaved., Tsvet. Met.*, 52 (1959).
10. A. Rabenau and H. Schulz, *J. Less-Common Metals 50*, 155–159 (1976).
11. P. F. Adams, P. Hubberstey, and R. J. Pulham, *J. Less-Common Metals 42*, 1–11 (1973).
12. C. C. Addison, B. M. Davies, R. J. Pulham, and D. P. Wallace, in: *The Alkali Metals*, Spec. Publ. No. 22, The Chemical Society, London, 1967, pp. 290–308.
13. P. I. Fedorov and M. T. Su, *Hua Hsueh Hsueh Pao* (J. Chinese Chem. Soc.) *23*, 30–39 (1957).
14. R. M. Yonco and M. I. Homa, *Trans. Am. Nucl. Soc. 32*, 270–271 (1979).
15. J. R. Weston, W. F. Calaway, R. M. Yonco, and V. A. Maroni, in: *2nd Intern. Conf. on Liquid Metal Technology in Energy Production* (J. M. Dahlke, Ed.), National Techn. Information Service, Springfield, Va., 1980 (CONF-800401-P2), Vol. 2, 20-1.
16. R. M. Yonco and M. I. Homa, *J. Nucl. Mater. 138*, 117–122 (1986).
17. R. Ainsley, A. P. Hartlib, P. M. Holroyd, and G. Long, *J. Nucl. Mater. 52*, 255–276 (1974).
18. B. Longson and A. W. Thorley, *J. Appl. Chem. 20*, 372–379 (1970).

19. F. Tepper and J. Grew, Report AFML-TR-64-327, 1964.
20. R. M. Yonco, E. Veleckis, and V. A. Maroni, *J. Nucl. Mater. 57*, 317–324 (1975).
21. E. Veleckis, K. E. Anderson, F. A. Cafasso, and H. M. Feder, in: *Proc. Intern. Conf. on Sodium Technology and Large Fast Reactor Design*, Report ANL-7520, Argonne, Ill., 1968, Part 1, pp. 295–298.
22. H. Schneider, in: *Material Behavior and Physical Chemistry in Liquid Metal Systems* (H. U. Borgstedt, Ed.), Plenum Press, New York, 1982, pp. 71–80.
23. C. Parmentier and F. Lievens, The Potentiometric Determination of Cyanide Carbon in Reactor Grade Sodium, Progress Report SCK/CEN Mol 1979.
24. H. U. Borgstedt, *Metall. 34*, 143–145 (1980).
25. K. Natesan and T. F. Kassner, *Metall. Trans. 4*, 2557–2566 (1973).
26. R. C. Asher and T. B. A. Kirstein, in: *Liquid Metals 1976*, Conf. Series No. 30, The Institute of Physics, Bristol and London, 1977, pp. 561–566.
27. F. J. Salzano, L. Newman, and M. R. Hobdell, *Nucl. Technol. 10*, 335–347 (1971).
28. T. Gnanasekaran, H. U. Borgstedt, and G. Frees, *Nucl. Technol. 59*, 165–169 (1982).
29. S. Rajendran Pillai and C. K. Mathews, *J. Nucl. Mater. 137*, 107–114 (1986).
30. W. E. Ruther, S. B. Skladzien, M. F. Roche, and J. W. Allen, *Nucl. Technol. 21*, 75–78 (1974).
31. R. J. Schlager, D. L. Olson, and W. L. Bradley, *Nucl. Technol. 27*, 439–441 (1975).
32. C. C. Gregg, USAEC Report UNC-5245, 1970.
33. H. U. Borgstedt and T. Gnanasekaran, in: *Analysis of Non-Metals in Metals* (G. Kraft, Ed.), de Gruyter, Berlin, 1981, pp. 455–460.
34. C. C. Addison, G. K. Creffield, P. Hubberstey, and R. J. Pulham, *J. Chem. Soc. (A)*, 1482–1487 (1969).
35. N. Rumbaut, F. Casteels, and M. Brabers, in: *Material Behavior and Physical Chemistry in Liquid Metal Systems* (H. U. Borgstedt, Eds.), Plenum Press, New York, 1982, pp. 437–444.

7

Hydrogen in Alkali Metals

The presence of hydrogen and its heavier isotopes deuterium and tritium in alkali metals is of technical importance for the application of the metals as working fluids in nuclear fission or fusion power-generating plants. The lithium–tritium system is of interest in fusion technology. Neutron reactions of lithium generate the tritium needed as nuclear fuel in the blanket of a fusion reactor. Tritium is the energy-generating material by means of the deuterium–tritium fusion process. The heavy isotope has to be extracted from the blanket fluid to be introduced into the plasma to replace the fuel consumed for the thermonuclear process.

The isotope tritium is among the fission products which occur in the fuel of fast neutron breeding reactors. It migrates by diffusion from the fuel elements into the core coolant. The sodium coolant has the capacity to dissolve tritium and to transfer it as a dissolved gas in the direction of the heat transfer. Thus, tritium may be transported through the sodium pipes into the sodium heat exchanger. The gas is able to escape the primary system by means of diffusion through the heat exchanger tubes, thus contaminating the secondary system. Activity enters systems outside the shielded areas in this way. Knowledge concerning the chemical properties of tritium solutions in liquid sodium and of the saturation concentrations enables us to prevent the migration of tritium into the secondary or even the tertiary systems.

Hydrogen also plays a role in the cooling systems of fast breeder reactors. Some hydrogen developed in the steam generators enters the secondary sodium circuits by means of diffusion through the steel tubes. The hydrogen is formed by oxidation reactions between the steel tubing and water vapor. The detection of hydrogen is an analytical problem, since fast and sensitive methods are necessary. The removal of hydrogen by means of trapping procedures is another problem. The solution of both problems is important for the safe and reliable operation of the reactor. An increase in hydrogen concentration in the secondary sodium indicates a leakage of the steam generator. Such events have to be detected as early as possible. The early signal indicating a steam generator leak enables the operating staff to shut down the reactor, before any serious sodium–water reaction may occur. The repair of small leaks in steam generator tubes can be easily performed.

Processes to generate hydrogen as an energy source may also be based on the sodium–hydrogen system. Such a process may be combined with a high-temperature reactor. The energy necessary for the electrolysis of water is lower at higher operating temperature of the electrochemical cell. In such an electrolytic device, the nickel cathode is connected to a sodium circuit. The permeability of the cathode to hydrogen developed at its surface allows the trapping of this element in the liquid sodium. At moderate temperatures sodium hydride is formed and dissolved. The solution of hydride in sodium moves through the circuit and decomposes in the region where sodium is heated to high temperatures by the heat of the reactor. The free hydrogen is separated from the circuit, and the sodium is returned into the process of hydrogen extraction. The process is claimed to consume less energy than conventional electrolysis.

The reason for extensive studies of the hydrogen–alkali metal systems is the technical importance of solutions of hydrogen and tritium in sodium and lithium. The reactions of the heavier alkali metals with hydrogen and its isotopes are less important, and only poor information on them is available.

7.1. Compounds of Hydrogen with Alkali Metals

The direct reaction of the alkali metals with gaseous hydrogen above 300°C leads to the formation of hydrides:

$$M + \tfrac{1}{2}H_2 \rightleftarrows MH \qquad (1)$$

Alkali hydrides are solid compounds of crystalline character. These saltlike compounds crystallize in the NaCl-type lattice, in which the hydride ions, H^-, take the places of the chloride ions. The apparent ionic radius of H^- is about 1.5 Å. Table 7.1 gives a survey of the properties of alkali hydrides, indicating the lower chemical stability of the compounds of the heavier elements of this group.

Alkali hydrides are also products of reactions of the molten metals

Table 7.1. Properties of Alkali Hydrides

Hydride	Molec. weight	Density ($g \cdot cm^{-3}$)	Melting temp. (°C)	Standard heat of formation ($kJ \cdot mol^{-1}$)	Lattice parameter (Å)
LiH	7.95	0.82	688	−91.0	4.093
NaH	24.00	0.92	800 dec.	−56.6	4.890
KH	40.11	1.47	430 dec.	−57.9	5.709
RbH	86.48	2.60	300 dec.	−47.4	
CsH	133.91	3.41	dec.	−49.9	

with water under protective atmospheres at sufficiently high temperatures. Another route to the formation of hydrides is the thermal decomposition of hydroxides in the presence of excess alkali metal:

$$2M + H_2O \rightarrow MOH + MH \tag{2}$$

and

$$MOH + 2M \rightarrow M_2O + MH \tag{3}$$

The alkali hydrides decompose at still higher temperatures. Lithium hydride is the only compound which can be heated to a temperature above the melting point without any decomposition. The hydrides develop a partial pressure of hydrogen in dilute solutions in the molten metals. This pressure may be applied to estimate the hydride concentration.

The hydrogen isotopes deuterium and tritium tend to form the same type of saltlike compounds. Lithium deuteride, for instance, has a lattice parameter $a = 4.0684$ Å.

7.2. Solubility of Hydrogen in Alkali Metals

Solid hydrides as well as gaseous hydrogen dissolve in liquid alkali metals. The solubility of hydrogen depends on the temperature, in a similar manner to the behavior of other elements, according to equation (4).

$$\ln[H]_s = A - \frac{B}{T} \tag{4}$$

The solubility of hydrogen in liquid lithium has been determined by Pulham *et al.*[1] by means of the electrical resistance method in the temperature range 523 to 775 K. The highest concentration they have measured is 5.68 mol %. The same group has estimated the solubility of deuterium in this alkali metal.[2]

Veleckis *et al.*[3] have also measured the solubility of lithium deuteride in the molten metal. They used a direct sampling method. The results of their study are closer to the solubility curve of hydride as determined at Nottingham University. Figure 7.1 compares the results of these measurements. The authors have not calculated a solubility equation, since they have not found an exactly linear function on the semilogarithmic scale. The published data fit the two solubility equations given below:

$$\text{Li/LiH: } \log[\text{H}]_s(\text{wppm}) = 8.6926 - \frac{2314.1}{T(\text{K})} \quad (5)$$

$$\text{Li/LiD: } \log[\text{D}]_s(\text{wppm}) = 8.8073 - \frac{2208}{T(\text{K})} \quad (6)$$

The solubility plots are compared with those of the other alkali hydrides in Fig. 7.2, where the unit weight parts per million is used because of its practical importance in alkali metal technology.

The tritium compound LiT has been found to have a solution behavior similar to that of the lighter isotopes.[4] A modified Sieverts apparatus has been used to measure its solubility in the molten metal, and the results are comparable to those fitted by equation (6).

Several methods have been used to measure the solubility of sodium hydride in liquid sodium. The group at Nottingham University[6] has used

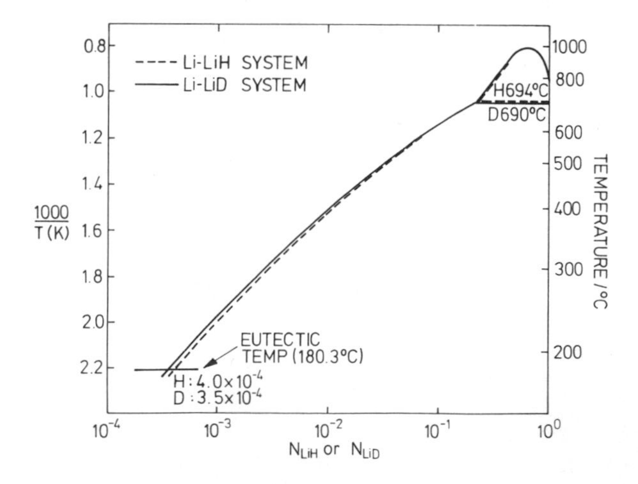

Figure 7.1. Phase diagrams of the Li–LiH and Li–LiD systems (after ref. 3; reproduced with the permission of Elsevier Sequoia, Oxford, UK).

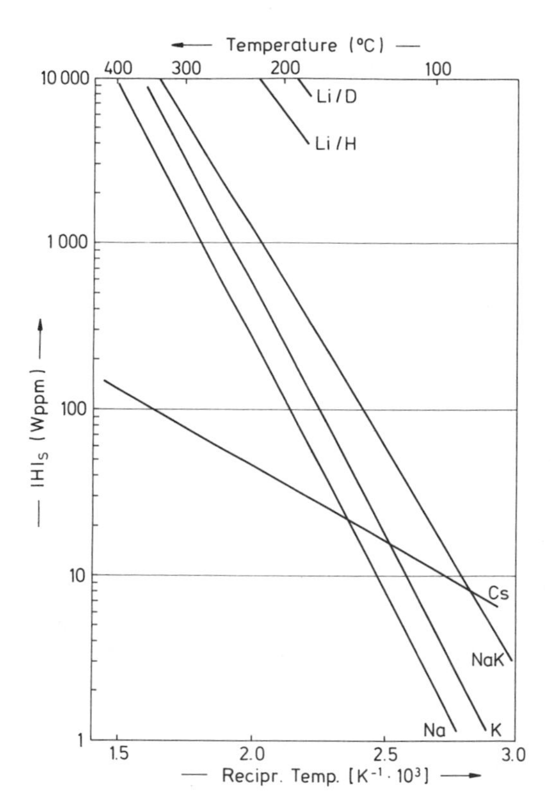

Figure 7.2. The solubility of hydrides in the molten alkali metals as a function of the reciprocal temperature according to equations (5) to (10).

the resistivity measurement method to establish the solubility curve in this system also. Another physicochemical method, a closed cell technique, has been used in the APDA laboratories in the United States.[7] The measurements are based on determinations of the hydrogen pressure developed by a closed volume of a well-defined sodium–sodium hydride mixture at a given temperature. This mixture, enclosed in a semipermeable sealed nickel capsule, develops a hydrogen partial pressure which is measured in the gas volume outside of the container.

The saturation concentration of hydrogen in sodium in the low-temperature region has been studied in a sodium loop.[8] In this study, the sodium was equilibrated with hydride precipitated in the cold trap. The loop was equipped with a diffusion-type hydrogen meter which monitored the hydrogen content in the sodium loop. A similar study using a different type of diffusion hydrogen meter produced nearly identical values of the saturation concentrations.[9] The solubility equation fitting the experimental results best is

$$\text{Na/NaH: } \log[\text{H}]_s(\text{wppm}) = 8.5184 - \frac{3051.3}{T(\text{K})} \quad (7)$$

All these studies have shown that liquid sodium has the capacity to dissolve considerably higher amounts of hydrogen at higher temperatures and higher pressures. At 1173 K and a pressure of 650 bar, for instance, the solubility of sodium hydride in the liquid metal has been reported to be about 19 mol %.[10]

The solubility of hydrogen as NaH or KH in the eutectic sodium–potassium alloy has also been determined because of its technical importance.[11,12] The experimental results are in accordance with the solubility equation

$$\text{NaK/(Na, K)H: } \log[\text{H}]_s(\text{wppm}) = 8.3271 - \frac{2624.6}{T(\text{K})} \quad (8)$$

Figure 7.2 shows that the eutectic dissolves considerably higher amounts of hydrogen than the pure alkali metals. This has to be considered when cold trapping of hydride from the eutectic melt at a temperature of ~25°C is discussed.

Studies of the potassium–hydrogen system have shown that it behaves similarly to the sodium–hydrogen system. The data[12] are used to formulate equation (9):

$$\text{K/KH: } \log[\text{H}]_s(\text{wppm}) = 8.6756 - \frac{2973.7}{T(\text{K})} \quad (9)$$

Solubility data for rubidium hydride in the molten metal have not yet been reported. However, some measurements in the cesium–cesium hydride system have been made.[13] They are in agreement with equation (10).

$$\text{Cs/CsH: } \log[\text{H}]_s(\text{wppm}) = 3.492 - \frac{914.6}{T(\text{K})} \quad (10)$$

This equation differs considerably from the solubility equations of the other alkali hydrides in the liquid metals. Therefore, equation (10) has to be considered with caution. The different slope of the CsH/Cs solubility plot is also obvious from Fig. 7.2.

The free enthalpies of solution of the alkali hydrides in molten alkali metals are calculated from equations (5) to (10). In Table 7.2 the values for the different alkali metal hydride solutions are compared.

Table 7.2. Free Enthalpies of Solution of Alkali Hydrides in Liquid Alkali Metals

Hydride	ΔH_{sol} (kJ · mol^{-1})
LiH/Li	43.6
LiD/Li	41.6
NaH/Na	57.6
(Na, K)H/NaK	49.5
KH/K	56.1
CsH/Cs	17.3

7.3. Hydrogen Pressure of Solutions of Hydrides in Alkali Metals

Solutions of hydrogen in alkali metals have a measurable hydrogen partial pressure which depends on the concentration of hydride and the temperature of the solution. Figure 7.3 shows the pressure of solutions of sodium hydride in sodium metal as a function of the concentration at a temperature of 677 K, as measured by Whittingham.(15) The diagram shows a nonlinear function and a plateau pressure above an upper value of the concentration. This concentration corresponds to the saturation point. The precipitation of sodium hydride is the reason that increases in the

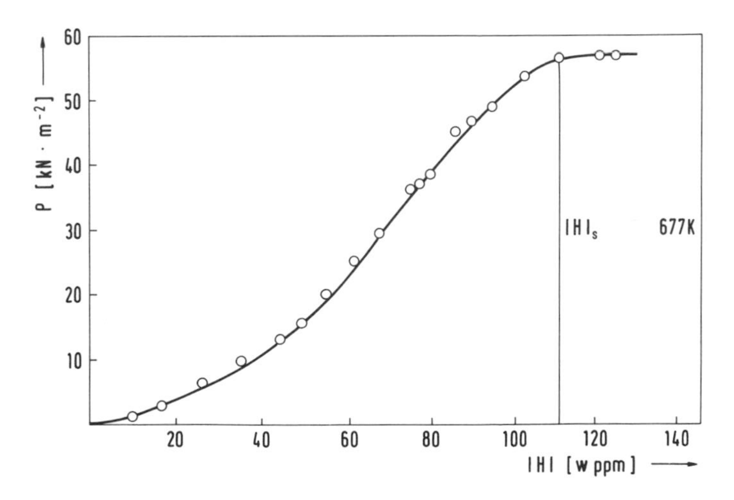

Figure 7.3. Hydrogen partial pressure of NaH solutions in liquid Na as a function of hydrogen concentration at 677 K (after ref. 15; reproduced with the permission of North Holland Publishing Co., Amsterdam, The Netherlands).

amount of hydride cannot generate higher levels of hydrogen pressure. If one plots the function shown in Fig. 7.3 against the square root of the partial pressure of hydrogen, the relationships become linear in accordance with Sieverts' law:

$$[\mathrm{H}] = K_s \cdot (p_{\mathrm{H}_2})^{1/2} \tag{11}$$

where [H] is the hydrogen concentration in wppm, p_{H_2} is the partial pressure of hydrogen in $\mathrm{N/m^2}$ in equilibrium with the liquid solution, and K_s is the Sieverts' constant. This constant depends on the temperature according to equation (12).

$$\ln K_s = \ln(T[\mathrm{H}]/p_{\mathrm{H}_2}^{1/2}) = A - \frac{B}{T} \tag{12}$$

Knowing the value of K_s of an individual alkali metal–hydrogen system enables us to use the partial pressures to estimate the concentrations of hydrogen in the molten alkali mctals. The partial pressures of hydrogen and its isotopes deuterium and tritium have been extensively studied in recent years. Thus, the older data have been revised and are now considered to be more or less unreliable. These most recent data are published in refs. 3–5 and 14.

The decomposition pressure data for LiH, LiD, and LiT in lithium have been measured below and above the monotectic temperature and $\ln P = A - (B/T)$ relationships for both regions have been estimated,[14] as is shown in Table 7.3. The constants A and B of equation (12), from which the Sieverts' constants can be calculated at temperatures of concern, are given for the temperature range above the monotectic temperature by

LiH: $A = -0.563$; $B = -6384$.
LiD: $A = -0.358$; $B = -5964$.
LiT: $A = -0.242$; $B = -5794$.

Table 7.3. Decomposition Pressure Data for the Li–LiH, Li–LiD, and Li–LiT Systems

System	Monotectic temperature (°C)	$\ln P = A - B/T$ Below P monotectic A	Below P monotectic $-B \times 10^3$	Above P monotectic A	Above P monotectic $-B \times 10^3$
Li–LiH	694	27.50	23.38	21.34	17.42
Li–LiD	690	28.13	23.59	21.15	16.87
Li–LiT	688	27.87	23.21	20.92	16.53

The thermodynamic data for the sodium–sodium hydride system have recently been reviewed by Whittingham.[15] This review shows that the data obtained so far for this system are in fair agreement. The Sieverts' constants, which are only slightly dependent on the temperature in the region between 600 and 687°C, are in the range of $K_s = 12.5$ wppm $\cdot$ $(\text{kN} \cdot \text{m}^{-2})^{-1/2}$ or $K_s = 4.6$ wppm $\cdot$ torr$^{-1/2}$, much higher than in the lithium–hydrogen system, thus indicating the higher decomposition

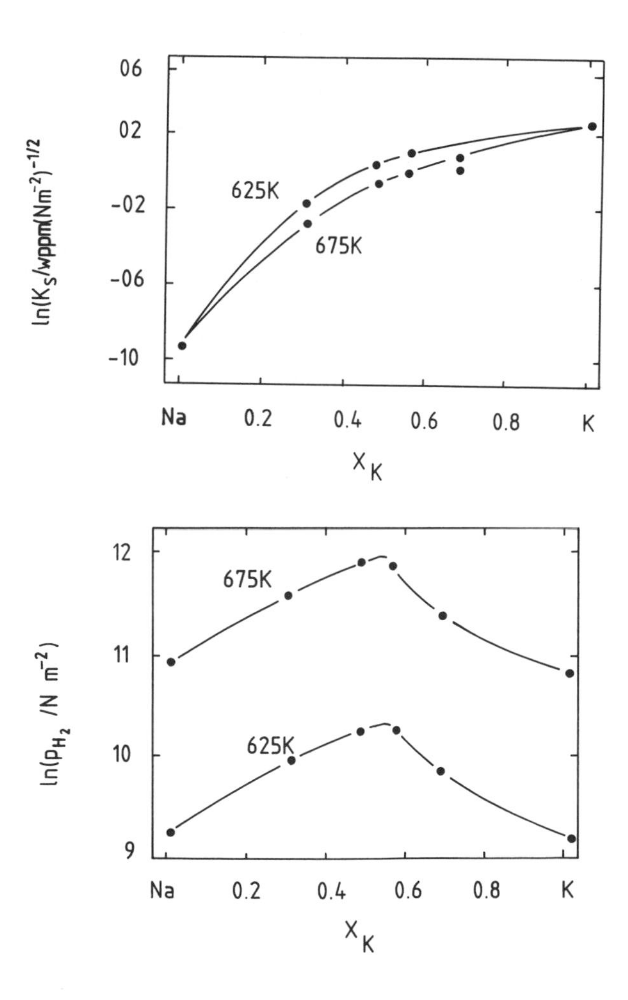

Figure 7.4. Sieverts' law constants ($\ln K_s$) and saturation hydrogen pressure at equilibrium ($\ln p^s_{H_2}$) vs. composition isotherms at 352 and 402°C for the NaK–Na(K)H system.

pressure of sodium hydride. The equation which fits best the relevant data given in ref. 15 is

$$\log P_s(\mathrm{N \cdot m^{-2}}) = 13.820 - \frac{6122}{T(\mathrm{K})} \tag{13}$$

Only three sets of data concerning the equilibrium partial pressures of hydrogen in potassium have been reported so far.[13,16,17] The results of these studies are in good agreement. Data based on the most extensive and consistent results[13] are recommended for application in alkali metal chemistry. Equation (14) relates these data to the temperature,

$$\ln p^s_{\mathrm{H_2}} = 31.12 - \frac{13{,}723}{T(\mathrm{K})} \tag{14}$$

Data for the NaK-Na(K)H system are shown in Figure 7.4. Data on equilibrium dissociation pressures of rubidium hydride are available.[16] Measurements in dilute solutions, however, have not been performed. The available data are summarized by equation (15).

$$\ln p^s_{\mathrm{H_2}} = 32.06 - \frac{13{,}079}{T(\mathrm{K})} \tag{15}$$

A study of the cesium–hydrogen system[19] has generated some partial pressure data at equilibrium as a function of the concentration of cesium hydride at two temperatures (300 and 370°C), and also some data on the cesium hydride dissociation pressure are available.[16] Since the data of the more recent study do not seem to be consistent with the data for the hydrides of the lighter alkali metals, as in the case of rubidium hydride, the dissociation pressure data for cesium hydride are applied to formulate equation (16).

$$\ln p^s_{\mathrm{H_2}} = 32.04 - \frac{13{,}585}{T(\mathrm{K})} \tag{16}$$

7.4. Analysis of Hydrogen in Alkali Metals

The determination of hydrogen in alkali metals is based on several methods such as vacuum extraction, amalgamation, or isotope dilution. Most recently, electrochemical methods and equilibration of hydrogen between alkali metals and a solid metal have also been introduced. Among the physical methods, the measurements of the electrical conductivity of alkali metal–alkali hydride mixtures are of importance. These methods have been applied for analyses of samples in the laboratory. Monitors based on some of the above-mentioned methods have been developed, however, and they may be applied to estimate concentrations of hydrogen (or its isotopes) in liquid metal loops, including the primary or secondary systems of fast sodium-cooled reactors.

The vacuum extraction method uses samples which are enclosed in sealed capsules of metals, such as iron or nickel, which are permeable to hydrogen. The capsules are heated to 700°C in a vacuum apparatus. Hydrogen which may be present as hydride or as hydroxide is completely removed from the sample by this procedure. The evolved hydrogen gas is measured in the evacuated volume by means of a sensitive pressure gauge or a gas chromatograph. The method published originally[20] is based on the conversion of hydrogen into water vapor by means of reaction with a layer of hot copper oxide and subsequent measurement of the decrease in pressure. The combination of vacuum extraction and gas chromatography is a very sensitive method, 5 μg in samples of 5 g being detected.

The isotope dilution method uses deuterium or tritium which is equilibrated with the hydrogen-containing alkali metals at high temperatures. The isotope ratio in the gas phase above the equilibrated solution—as determined by means of a mass spectrometer—is used to calculate the original hydrogen content of the metal sample. Sensitivity and accuracy are, however, not better than with the vacuum extraction method. Thus, there is no reason to apply this more complicated and expensive method.

The amalgamation method[22] is reported to be sensitive to 0.1 wppm and is applicable for the determination of hydrogen even in the presence of oxygen. Hydride and hydroxide can be differentiated by means of control of the amalgamation temperature. The hydride which remains after the dissolution of the alkali metal has to be decomposed. Thermal or hydrolytical methods are available for the liberation of the hydrogen gas for the measurement.

The principle of the vacuum extraction method is also applied in diffusion hydrogen meters which have been developed as monitors for application in sodium circuits. Figure 7.5 shows a particular type of such a meter. A metallic membrane permeable to hydrogen and corrosion resistant against alkali metals (nickel) is immersed in the liquid metal loop.[8] The membrane separates the liquid metal from the gas volume which serves as the device for the gas analysis. This volume is connected to an ion getter pump which may be used in the dynamic mode. In this operation mode the hydrogen diffusing into the vacuum is estimated from the current of the ion getter pump itself. The meter is very sensitive to

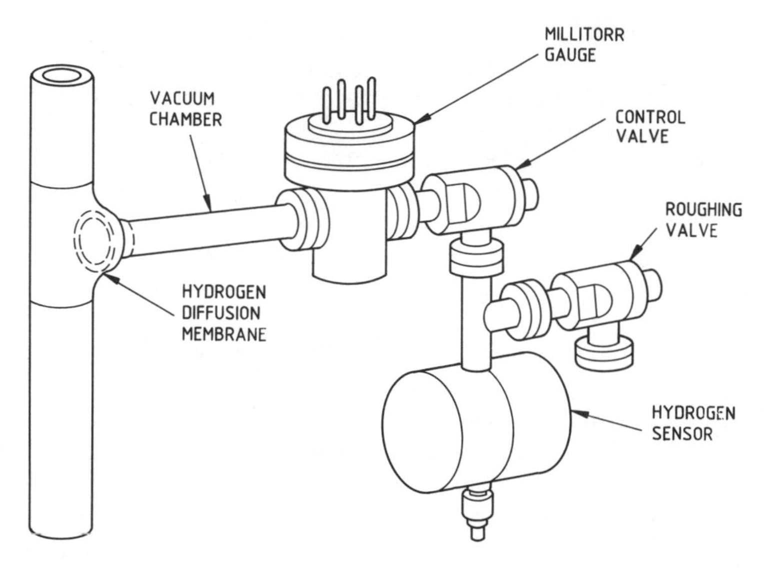

Figure 7.5. Diffusion hydrogen meter with capacity for dynamic-mode measurements and equilibrium-mode measurements for determination of absolute hydrogen partial pressures (Varian).

changes in the concentration of hydrogen in the liquid metal, if it is applied in this mode. The operation in the static mode offers the possibility of determining the equilibrium hydrogen pressure of the alkali metal. The control valve of the meter has to be closed for this purpose. Thus, hydrogen diffuses into the evacuated cell up to the equilibrium pressure. This equilibrium is reached after 10 to 15 minutes. It is determined by means of the millitorr gauge connected to the upper part of the vacuum chamber. Experiments with a meter connected to a sodium loop have shown that the hydrogen partial pressures are controlled by the cold-trap temperature of such loops. The application of such diffusion hydrogen meters makes any additional analytical equipment unnecessary.

The diffusion meters are sensitive and specific. At concentration levels of 0.1 ppm hydrogen in sodium, changes of 0.02 ppm give a suitable change in pressure or ion getter pump current values.[9] Diffusion hydrogen meters may also be equipped with gas chromatographic or mass spectrometric units to measure the hydrogen concentration on the gas side of the instrument.

The electrochemical cell for hydrogen determination in alkali metals described by Smith[21] is based on a molten salt electrolyte placed in a cup of iron or nickel, which is permeable to hydrogen. The molten electrolyte consists of a mixture of calcium chloride and calcium hydride, the concentration of which is less than 15 mol %. The cell develops a voltage depending on the temperature T and the partial pressure of hydrogen p_{H_x} in relation to the reference partial pressure of hydrogen p_{H_r}.

$$E = E_0 + A \cdot \frac{RT}{2F} \cdot \ln \frac{p_{H_x}}{p_{H_r}} \tag{17}$$

E_0 should be zero, but has in fact a value between zero and -510 mV, the value of A is a fraction of the theoretical EMF of the cell, depending on the temperature and the hydrogen potential, R is the gas constant, and F the Faraday constant. The cell has to be calibrated in order to compensate for the difference between the cell voltage and the theoretical EMF. In the temperature range from 500 to 600°C, the cell shown in Fig. 7.6 has been found to show values close to theory even in the concentration region of 1 wppm hydrogen in sodium. A more recently developed version uses lithium/lithium hydride as the reference system,[22] thus leading to the electrochemical cell

$$\text{Na, NaH} \,\|\, \text{Fe, CaCl}_2\text{, 15 mol \% CaH}_2\text{, Fe} \,\|\, \text{Li, LiH}_{sat}$$

This type of hydrogen meter has successfully been used to detect water leakages in sodium loops. Electrochemical hydrogen meters in this final version have the capacity to estimate absolute values of chemical activities of dissolved hydrogen in several alkali metals. They offer an opportunity to reestablish the saturation concentration values by means of an independent method.[23]

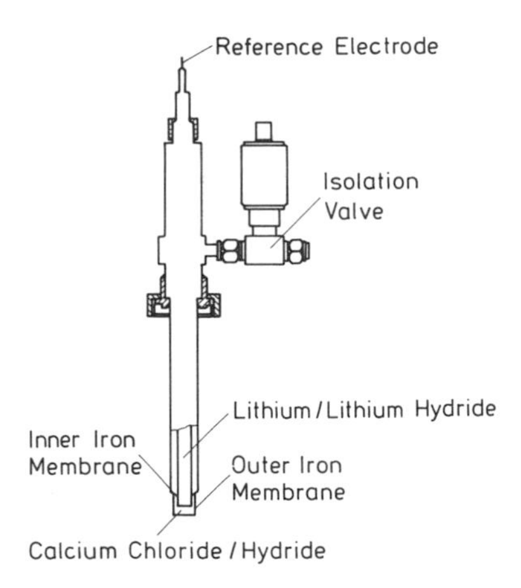

Figure 7.6. Principle and design of the electrochemical hydrogen meter for application in sodium circuits (after ref. 21).

Additionally, it can be said that the equilibration of solid metals or alloys offers an indirect method of estimating hydrogen chemical activities. The solid metals and the alkali melts have to be heated together to reach the equilibrium distribution of the nonmetal between the two metallic phases. Hydrogen may be estimated in the solid material by means of the vacuum fusion method. The measured concentration has to be converted into chemical activity. This value is the same as in the alkali melt due to chemical equilibrium of the distribution reaction. Metals used for equilibration of hydrogen between themselves and lithium or sodium are pure metals of the rare earths group, mainly scandium.

References

1. P. F. Adams, M. G. Down, P. Hubberstey, and R. J. Pulham, *J. Less-Common Metals 42*, 325–334 (1975).
2. P. F. Adams, P. Hubberstey, R. J. Pulham, and A. E. Thunder, *J. Less-Common Metals 46*, 285–289 (1976).
3. E. Veleckis, R. M. Yonco, and V. A. Maroni, *J. Less-Common Metals 55*, 85–92 (1977).
4. V. A. Maroni, W. F. Calaway, E. Veleckis, and R. M. Yonco, in: *Intern. Conf. on Liquid Metal Technology in Energy Production* (M. H. Cooper, Ed.), National Techn. Information Service, Springfield, Va., 1976, (CONF-760503-P1), Vol. 1, pp. 437–445.
5. F. J. Smith, J. B. Talbot, J. F. Laud, and J. T. Bell, in: *Radiation Effects and Tritium Technology for Fusion Reactors* (J. S. Watson and F. W. Wiffen, Eds.), National Techn. Information Service, Springfield, Va., 1976, (CONF-750989), Vol. 2, pp. 539–553.
6. C. C. Addison, P. Hubberstey, J. Oliver, R. J. Pulham, and P. A. Simm, *J. Less-Common Metals 61*, 123–132 (1978).
7. S. A. Meacham, E. F. Hill, and A. A. Gordus, The Solubility of Hydrogen in Sodium, USAEC Report APDA-241, 1970.
8. D. R. Vissers, J. T. Holmes, L. G. Bartholme, and P. A. Nelson, *Nucl. Technol. 21*, 235–244 (1974).
9. N. P. Bhat and H. U. Borgstedt, *Atomkernenergie 35*, 170–174 (1980).
10. W. Klostermeier and E. U. Franck, *Ber. Bunsenges. Phys. Chem. 86*, 606–612 (1982).
11. E. L. Compère and J. E. Savolainen, *Nucl. Sci. Eng. 28*, 325–337 (1967).
12. M. N. Ivanovskij, M. N. Arnol'dov, V. A. Morovzov, T. I. Moiseeva, and S. S. Pletenez, *Izv. Akad. Nauk. SSSR, Metal. 1*, 201–205 (1980).
13. M. N. Arnol'dov, Y. N. Bogdanov, M. N. Ivanovskij, and V. A. Morojov, *Izv. Akad. Nauk. SSSR, Metal. 4*, 30–32 (1976).
14. E. Veleckis, *J. Nucl. Mater. 79*, 20–27 (1979).
15. A. C. Whittingham, *J. Nucl. Mater. 60*, 119–131 (1976).
16. A. A. Hérold, *Ann. Chim. 9*, 536–581 (1951).
17. A. R. Kurbanov, A. Badalov, and U. Mirsaidov, *Dokl. Akad. Nauk. Tadzh. SSR 23*, 83–86 (1980).
18. L. P. Pepkowitz and E. R. Proud, *Anal. Chem. 21*, 1000–1003 (1949).
19. B. D. Holt, *Anal. Chem. 31*, 51–54 (1951).
20. G. Naud, Contribution à l'étude des impuretés hydrogénées et oxygénées dans le sodium liquide, Rapport CEA-R2583, 1964.

21. C. A. Smith, in: *Liquid Alkali Metals*, British Nuclear Energy Society, London, 1974, pp. 101–106.
22. G. J. Licina, P. Roy and C. A. Smith, in: *Material Behavior and Physical Chemistry in Liquid Metal Systems* (H. U. Borgstedt, Ed.), Plenum Press, New York, 1982, pp. 297–307.
23. V. Ganesan, T. Gnanasekaran, R. Sridharan, G. Periaswami, and C. K. Mathews, in: *Liquid Metal Engineering and Technology*, British Nuclear Energy Society, London, 1984, Vol. 1, pp. 369–373.

8

Other Nonmetals

In the last three chapters we have considered in detail the interactions of the most important nonmetals, viz., carbon, nitrogen, oxygen, and hydrogen, with liquid alkali metals. The solubilities of the remaining nonmetallic elements and the types of compounds that precipitate at saturation are discussed in this chapter. The data available in the literature are too meager to present a comprehensive picture. From the technological point of view there is more interest in nonmetallic elements which (1) form minor constituents of structural materials (e.g., P, Si), (2) are produced in fission (e.g., Te, I), or (3) can cause corrosion (e.g., halogens). In addition, there is considerable interest in rare gases as they are encountered in cover gases (Ar, He) and in fission products (Kr, Xe). The solubilities of these nonmetals depend to a large extent on the properties of the precipitating phase. In general, there is a trend of increasing solubility with increasing metallic character of the solute atom.

8.1. Solubilities of Rare Gases in Alkali Metals

The solubility of a noble gas in a liquid alkali metal depends on the gas pressure

$$S = K_H \cdot P \tag{1}$$

where S is the solubility in atom fraction, P is the pressure, and K_H is the Henry's law constant. Gas solubilities are, therefore, conveniently expressed in terms of the Henry's law constant in units of atom fraction solute in solution per unit gas pressure.

Solubilities of rare gases in liquid alkali metals have been recently reviewed by Hubberstey.[1] Earlier reviews of Reed and Droher,[2] Thormeier,[3] and Foust[4] are essentially confined to sodium. As most of the experimental measurements have been carried out on sodium, we will first discuss the solubilities of rare gases in sodium and then consider the information available on other alkali metals.

8.1.1. Solubilities of Rare Gases in Sodium

Though a few measurements were carried out in the sixties (see refs. 2, 3, and 4), the more reliable data appeared in the seventies. From the Argonne National Laboratory, Veleckis *et al.*[5,6] reported the solubilities of helium, argon, and xenon and Bloomquist *et al.*[7] the solubility of krypton. Thormeier[8] measured the solubilities of helium and argon. Shimijima and Miyaji[9] reported the solubility of neon. The ANL group has also proposed a solution model to calculate the solubilities of noble gases in liquid sodium.

The methods used by these workers for solubility measurements were very similar. Liquid sodium was saturated with the gas at a well-defined temperature and pressure. From a known quantity of the saturated solution, the gas was stripped and quantitatively assayed. The differences between the different groups were in the techniques used for equilibration, stripping, and assay. The ANL group bubbled the gas through the liquid to

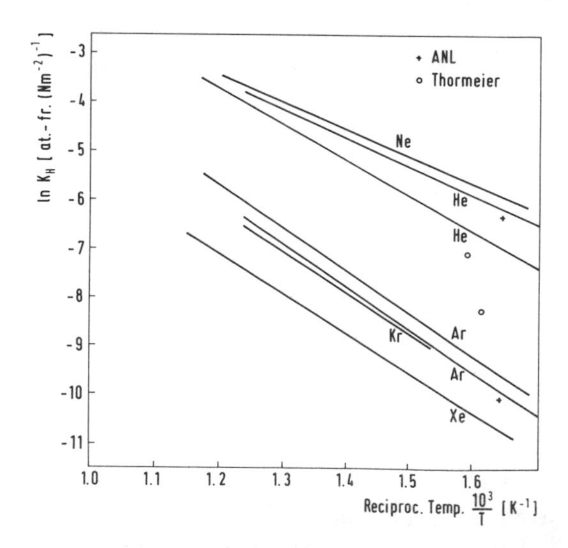

Figure 8.1. Variation of Henry's law constant for noble gases in liquid sodium (after ref. 1; reproduced with the permission of Blackwell Scientific Publications, Oxford, UK).

achieve saturation, sparging by another inert gas for stripping, and mass spectrometry and gas chromatography for assay. Thormeier pumped liquid sodium into a tank containing the gas for equilibration, evacuation for stripping, and volumetry for assay. No experimental details were given by the Japanese workers.

All the solubility data provided by the above studies are compared in Fig. 8.1 which is taken from ref. 1. A general trend of decreasing solubility with increasing size of the inert gas atom is observed. The solubility of neon does not appear to fit into this general trend. It must, however, be noted that the neon data were reported by an independent group of workers. Nevertheless, it appears that the solubility plots for helium and neon are close to each other, as are those of argon and krypton. Cafasso *et al.*[10] have pointed out that this is consistent with their model. The hard-sphere diameter of krypton (0.36 nm) is only 5.6% larger than that of argon but its polarizability is 34% greater, thus allowing for its better accomodation in the liquid metal as a result of electrostatic attraction. A similar case prevails with the helium–neon pair. The temperature dependence of the Henry's law constant can be expressed in the form

$$\ln K_H = A - \frac{B}{T} \tag{2}$$

The constants A and B given in Table 8.1 are taken from ref. 1. The units of K_H are atom fraction per newton per square meter.

8.1.2. Solubilities of Rare Gas in Other Alkali Metals

The solubility of helium has been measured by Slotnick and Kapelner[11] in both potassium and lithium. For lithium, measurements were carried out in the temperature range of 922 to 1144 K at a pressure of

Table 8.1. Solubilities: Constants of the K_H vs. T^{-1} Expressions

Solvent	Solute	A	B	Temperature range (K)
Lithium	Helium	−5.301	1984	922–1144
Sodium	Helium	4.247	6523	605–825
Sodium	Neon	4.461	6599	625–850
Sodium	Argon	5.562	9719	605–825
Sodium	Krypton	5.970	10072	675–825
Sodium	Xenon	4.549	9774	625–875
Potassium	Helium	2.291	2246	755–977

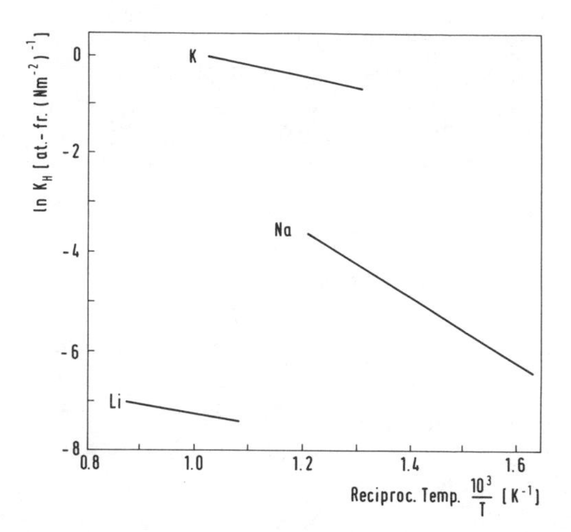

Figure 8.2. Variation of Henry's law constant for solutions of He in Li, Na, and K with temperature.

2.38 bar. In the case of potassium the measurement range was 755 to 977 K and 1 to 3 bar pressure. The data are described by the equations

$$\log(\text{mole He/mole Li/bar}) = -7.33 - \frac{840}{T(\text{K})} \quad (3)$$

$$\log(\text{mole He/mole K/bar}) = -3.30 - \frac{1550}{T(\text{K})} \quad (4)$$

These data are plotted in Fig. 8.2 along with the solubility of helium in sodium. There is a general trend of increasing solubility with the size of the alkali metal atom. However, the slopes are quite different. Hubberstey has pointed out that since the Na–He data are more reliable, the Li–He and K–He data must be considered preliminary until confirmed by an independent study.

There are no reported measurements of the solubilities of other inert gases in lithium or potassium, nor of the solubilities of any of the rare gases in the sodium–potassium eutectic and the heavy alkali metals cesium and rubidium.

8.2. Solubilities of Halogens in Alkali Metals

The halides of alkali metals of the type MX are well known, and these are the solid phases that would precipitate from solutions of halogens in alkali metals. Information on the solubilities of halogens, especially at relatively low temperature is, however, sparse.

During the period 1955 to 1962 Bredig and his coworkers published several papers on their studies on the miscibility of alkali metals with their molten halides.[12–17] Primarily they followed the cooling curves of metal–metal halide mixtures in order to delineate phase transition temperatures. They observed liquid immiscibility gaps in most cases. A typical set of results from their work[14] is shown in Fig. 8.3 covering the KX–K systems (X = F, Cl, Br, I). Barker[18] has recently collated Bredig's solubility data and given them in the form of a graph which is reproduced in Fig. 8.4. The general trend of increasing solubilities when going from sodium to cesium is obvious. Less clearly discernible is the trend of increasing solubilities as one goes down the halogen group. Bredig's measurements mainly cover the high-temperature region and there is very little data on the low-temperature region where solubilities are low but nevertheless interesting in alkali metal applications.

Miscibility of lithium with lithium chloride was investigated by Nakajima *et al.*[19] who used differential thermal analysis for high-temperature studies and sampling followed by chemical analysis for low-temperature solubilities. Konovalov *et al.*[20] also studied dilute solutions using

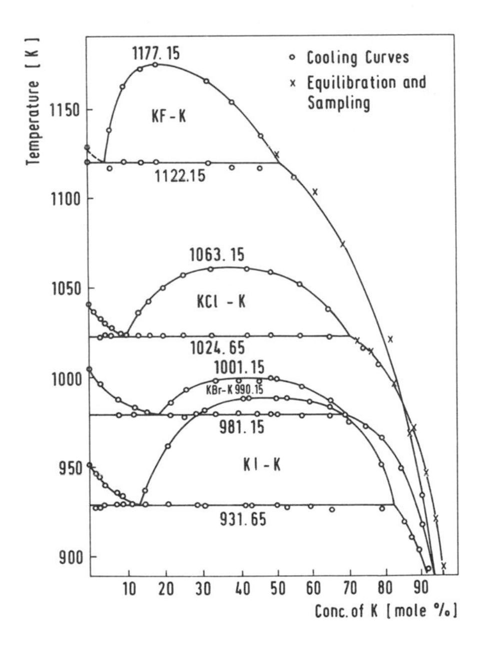

Figure 8.3. Temperature–composition phase diagrams of the KX–K systems (X = halogen) (after ref. 14).

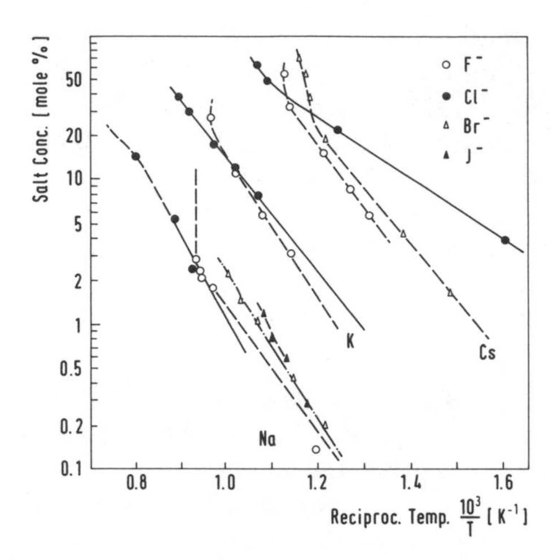

Figure 8.4. Solubility of solid alkali halides in alkali metals (after ref. 18; reproduced with the permission of Blackwell Scientific Publications, Oxford, UK).

the procedure of equilibration and sampling followed by neutron activation analysis. The two sets of data differ by more than an order of magnitude. While Nakajima's solubilities range from 0.007 mol % at 567 K to 0.23 mol % at 923 K, those of Konovalov range from 0.001 at 471 K to 0.011 mol % at 873 K.

Similar discrepancy exists in the limited low-temperature data available on sodium as well. Allen[21] measured the solubilities of NaBr and NaI in the temperature range 423 to 673 K in sodium containing 10 and 20 wppm of oxygen. He used labeled NaBr and NaI for equilibration in sealed capsules and in recirculating loops. Both gamma spectroscopy and chemical analysis were used to follow the halides. Clough[22] deduced bromide solubility in sodium by following the disappearance of a radioactive NaBr deposit on batchwise equilibration with known quantities of sodium. Barker[18] has combined Allen's and Clough's data to give the following expression for bromine solubility in sodium

$$\log S\,(\text{wppm NaBr}) = 9.00 - \frac{5100 \pm 240}{T(\text{K})} \tag{5}$$

Similarly, Allen's data were fitted to give the following solubility expression for iodine in sodium.

$$\log S\,(\text{wppm NaI}) = 8.72 - \frac{4650 \pm 140}{T(\text{K})} \tag{6}$$

Except for Bredig's data, which are not sufficient to derive solubility expressions, there are no other reported investigations of halide solubilities in other alkali metals.

8.3. Group VIA Elements

The solubility of oxygen in alkali metals has already been discussed in Chapter 5. As in the case of oxygen, the compound that coexists with the liquid metal is M_2X where $X = S$, Se, Te. These compounds are well known.

The sodium–sulfur system is characterized by the existence of several polysulfides. The phase diagram of the Na–S system has been reported,[23] though details at the metal-rich end are not very clear. As the Na–S system is of interest in sodium–sulfur batteries, the phase diagram is given in Fig. 8.5. There are no reliable solubility data reported for sulfur in sodium.

Polyselenides and polytellurides of alkali metals are also known, but their properties and phase relationships are outside the scope of this book.

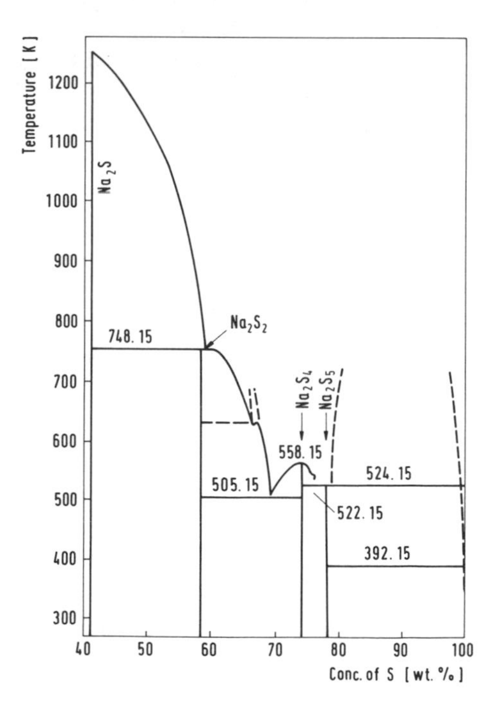

Figure 8.5. Phase diagram of the Na_2–S system.

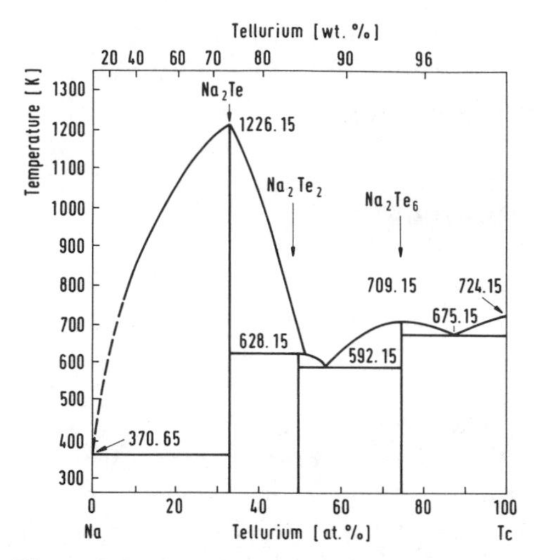

Figure 8.6. Phase diagram of the Na–Te system.

Alkali metal–selenium and alkali metal–tellurium phase diagrams are available in handbooks and they suggest that selenium and, more so, tellurium have significant solubilities in alkali metals at high temperatures. As an example, the sodium–tellurium phase diagram based on Hansen and Anderko[24] is given in Fig. 8.6. Considering the experimental techniques used, solubility data derived from these phase diagrams cannot be considered reliable.

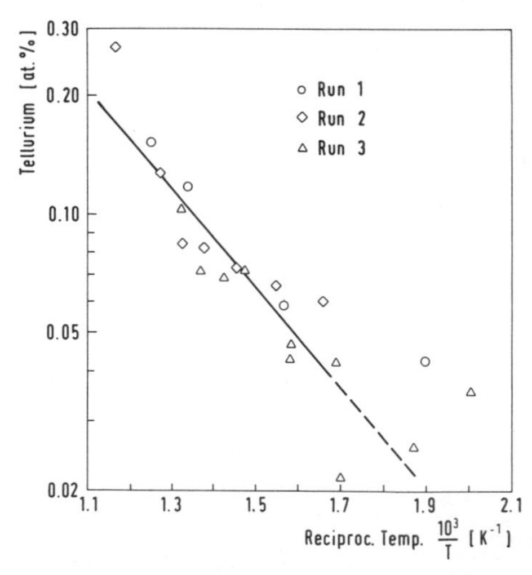

Figure 8.7. Solubility of tellurium in liquid sodium (after ref. 25; reproduced with the permission of North Holland Publishing Comp., Amsterdam, The Netherlands).

Walker and Pratt[25] measured the solubility of tellurium in sodium using an equilibration-and-sampling method. Their data are shown in Fig. 8.7. The solubility increases from about 0.03 at % at 495 K to 0.26 at % at 845 K. The solubility expression given by these authors is

$$\log S\,(\text{at \% Te}) = 0.7501 - \frac{1281.3}{T(\text{K})} \tag{7}$$

No reliable solubility data are available for other systems.

8.4. Group VA Elements

As we have seen in Chapter 6, the compound of nitrogen that coexists with liquid alkali metals is M_3N. Similar compounds of the formula M_3X (X = P, As, Sb, Bi) are known for all alkali metals. There are several compounds which are richer in X in the M–X system but they are outside the scope of the discussion here.

As phosphorus is a constituent of steels, its solubility in alkali metals is of interest. However, there are no reported studies of its solubility in these liquids. The same is the case with arsenic. The phase diagram of the Na–Sb system found in handbooks[24] indicates high solubility of antimony in sodium. However, Claar[26] has quoted Lamprecht's results to show that

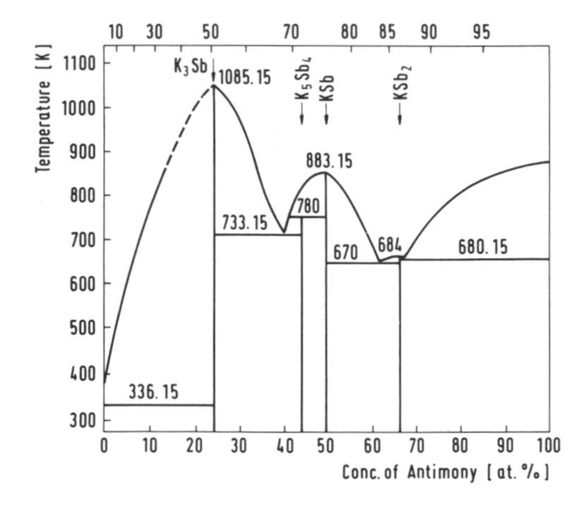

Figure 8.8. Phase diagram of the K–Sb system.

the solubility is only 17 wppm at 456 K and 1082 wppm at 823 K. He has recommended the following solubility expression:

$$\log S\,(\text{wppm Sb}) = 5.28 - \frac{1839}{T(\text{K})} \tag{8}$$

Phase diagrams of some of the higher alkali metals with antimony suggest considerable solubility. Examples are potassium–antimony[27] and cesium–antimony[28] systems. Figure 8.8 shows the phase diagram of the potassium–antimony system adopted from ref. 27. As expected, the solubility is higher when the metallic character of the solute element is higher. Thus, bismuth would be expected to have a good solubility in alkali metals but this metallic element is considered in the next chapter.

8.5. Group IVA Elements

In view of the importance of the chemistry of carbon in liquid alkali metals it has been treated separately in Chapter 6. The next element in the group, viz., silicon, is also of technological interest inasmuch as it is a minor constituent of steels. The solubilities of silicon and germanium are briefly considered in this section. The remaining elements of the group, being metals, come within the purview of the next chapter.

In the case of lithium, the precipitating phase is $Li_{22}M_5$ (M = Si, Ge), though Li_2C_2 is the carbon compound that exists in equilibrium with the liquid metal. In the case of sodium, the precipitating phase is NaM and here again carbon is the exception. Similar behavior is observed with the higher alkali metals.

Solubilities of silicon and germanium in lithium have been investigated by Fedorov *et al.*[29] and by Hubberstey *et al.*[30] The latter authors, who used an equilibrium resistivity method, have given the following solubility expressions:

$$500\text{–}700\text{ K: } \ln S\,(\text{mole fraction Si}) = 5.548 - \frac{6775}{T(\text{K})} \tag{9}$$

$$530\text{–}715\text{ K: } \ln S\,(\text{mole fraction Ge}) = 5.459 - \frac{6630}{T(\text{K})} \tag{10}$$

Hubberstey and Pulham[31] have used the same resistivity technique to measure the solubility of germanium in sodium. Their solubility data are shown in Fig. 8.9.

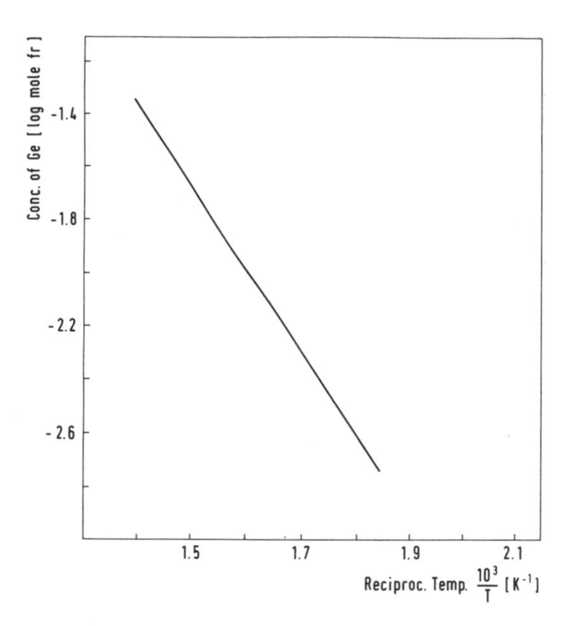

Figure 8.9. Solubility of germanium in sodium (after ref. 31).

There are no reports concerning the solubility of silicon in sodium, nor have the solubilities of these elements in the heavier alkali metals been reported.

References

1. P. Hubberstey, in: *Handbook of Thermodynamic and Transport Properties of Alkali Metals* (R. W. Ohse, Ed.), Blackwell Scientific Publications, Oxford, 1985, p. 907.
2. E. L. Reed and J. J. Droher, USAEC Report LMEC-69-36, 1970.
3. K. Thormeier, Report KfK-1166, Karlsruhe Nuclear Center, 1970.
4. O. J. Foust (Ed.), *Sodium Na–K Engineering Handbook*, Vol. 1, Gordon and Breach, New York, 1972.
5. E. Veleckis, S. K. Dhar, F. A. Cafasso, and H. M. Feder, *J. Phys. Chem. 75*, 2832–2838 (1971).
6. E. Veleckis, F. A. Cafasso, and H. M. Feder, *J. Chem. Eng. Data 21*, 75–76 (1976).
7. R. A. Bloomquist, F. A. Cafasso, and H. M. Feder, *J. Nucl. Mater. 59*, 199–200 (1976).
8. K. Thormeier, *Atomkernenergie 14*, 449 (1969); *Nucl. Eng. Des. 14*, 69–82 (1970).
9. H. Shimijima and N. Miyaji, *J. Nucl. Sci. Technol. (Japan) 12*, 658 (1975).
10. F. A. Cafasso, E. Veleckis, H. M. Feder, and H. C. Schnyders, in: *Intern. Conf. on Liquid Metal Technology in Energy Production* (M. H. Cooper, Ed.), National Techn. Information Service, Springfield, Va., 1976, (CONF-760503-P2), Vol. 2, pp. 619–622.
11. H. Slotnick and S. M. Kapelner, USAEC Report PWAC-380, 1965.
12. M. A. Bredig, H. R. Bronstein, and W. T. Smith, *J. Am. Chem. Soc. 77*, 1454–1458 (1955).

13. M. A. Bredig, J. W. Johnson, and W. T. Smith, *J. Am. Chem. Soc. 77*, 307–312 (1955).
14. J. W. Johnson and M. A. Bredig, *J. Phys. Chem. 62*, 604–607 (1958).
15. M. A. Bredig and H. R. Bronstein, *J. Phys. Chem. 64*, 64–67 (1960).
16. M. A. Bredig and J. W. Johnson, *J. Phys. Chem. 64*, 1899–1900 (1960).
17. A. S. Dworkin, H. R. Bronstein, and M. A. Bredig, *J. Phys. Chem. 66*, 572–573 (1962).
18. M. G. Barker, in: *Handbook of Thermodynamic and Transport Properties of Alkali Metals* (R. W. Ohse, Ed.), Blackwell Scientific Publications, Oxford, 1985, p. 905.
19. T. Nakajima, R. Minami, K. Nakanishi, and N. Watanabe, *Bull. Chem. Soc. Japan 47*, 2071–2072 (1974).
20. E. E. Konovalov, N. I. Seliverstov, and V. P. Emelyanov, *Izv. Akad. Nauk. SSSR, Metal. 3*, 109–112 (1968).
21. G. C. Allen, in: *Liquid Alkali Metals*, British Nuclear Energy Society, London, 1973, pp. 159–164.
22. W. S. Clough, *J. Nucl. Energy 25*, 417–423 (1971).
23. Gmelin: *Handbuch der Anorganischen Chemie*, System-Nummer 21 Natrium, Verlag Chemie, Weinheim, New York, 1967.
24. M. Hansen and K. Anderko, *Constitution of Binary Alloys*, McGraw-Hill, New York, 1958.
25. R. A. Walker and J. N. Pratt, *J. Nucl. Mater. 34*, 165–173 (1970).
26. T. D. Claar, *Reactor Technol. 13*, 124–146 (1970).
27. R. P. Elliott, *Constitution of Binary Alloys*, First supplement, McGraw-Hill, New York, 1965.
28. F. W. Dorn and W. Klemm, *Z. Anorg. Allg. Chem. 309*, 189–203 (1961).
29. P. I. Fedorov and A. A. Joffee, *Izv. Vyssh. Ucheb. Zaved., Tsvet. Met.*, No. 6, 127–131 (1961); P. I. Fedorov and V. A. Molochko, *Izv. Akad. Nauk. SSSR, Neorg. Mater. 2*, 1614–1616 (1966).
30. P. Hubberstey, A. T. Dadd, and P. G. Roberts, in: *Material Behavior and Physical Chemistry in Liquid Metal Systems* (H. U. Borgstedt, Ed.), Plenum Press, New York, 1982, pp. 445–454.
31. P. Hubberstey and R. J. Pulham, *J. Chem. Soc., Dalton Trans.*, 1541–1544 (1974).

9

Solubility of Metallic Elements in Liquid Alkali Metals

Alkali metals come in contact with various metals and alloys during their manufacture, handling, and applications. Therefore, the interaction of alkali metals with other metals is a matter of considerable importance. In general we find that group A (*s* and *p* block) and group B (*d* and *f* block) elements of the periodic table react differently with alkali metals. Many group A elements form intermetallic compounds with alkali metals. On the other hand, group B elements react much less with alkali metals. Thus transition metals, including inner transition metals (lanthanides and actinides), show very little solubility in liquid alkali metals. Coinage metals, platinum group metals, zinc, cadmium, and mercury are, however, exceptions. Therefore, the container materials for the alkali metals are usually made of transition metals or their alloys. Among the alkali metals the first member, lithium, behaves somewhat differently from the others. Thus the number of intermetallic compounds known in lithium systems is more than those known in other systems. Even transition metals which are virtually insoluble in sodium and potassium tend to dissolve in lithium. For example, manganese which dissolves to the extent of only 0.8 wppm in sodium at 700 K is soluble to the extent of 0.03 wt% in lithium at the same temperature. Typical solubilities of the constituents of stainless steels are compared in Table 9.1.

Solubilities of metals in liquid alkali metals are often influenced by the impurity content. Dissolved nitrogen in lithium and oxygen in other alkali metals particularly have a strong effect on solubility. The oxygen effect, for example, is very important for metals whose oxides are stable compared to the corresponding alkali metal oxide. Ternary oxygen compounds such as

Table 9.1. Solubilities at 700 K of Constituents of Stainless Steels in Alkali Metals

Element	Solubilities (in wppm)		
	Li	Na	K
Ni	87.50	0.67	0.31
Cr	1.38	0.0003	0.02
Fe	1.41	0.07	0.24
Mn	29.0	0.84	—

$NaCrO_2$ and Na_4FeO_3 play a role in the corrosion of steels by liquid sodium. Thus, "true" solubility in an alkali metal must be measured after ensuring its purity. On the other hand, dissolution rates at practical purity levels are important from the point of view of technological applications of alkali metals.

Different techniques have been employed for measuring solubilities of metals in alkali metals. The method employed depends on the magnitude of the solubility. Differential thermal analysis, a versatile tool in establishing phase diagrams, does not find much application in solubility measurement in alkali metals which exhibit a steep hypereutectic liquidus. The resistivity technique has been employed when solubility is low. However, the lower limit of measurement by this technique is a few thousand ppm of solute material. When the solubility is meager equilibration followed by concentration measurement by various trace analytical procedures has been employed.

The resistivity technique is based on the fact that the resistivity of a metal increases linearly with increasing concentration of the solute and that the resistivity coefficient is characteristic of the solute. The resistance of a liquid metal column in a stainless steel capillary (typically, 1.5 mm diameter and 350 mm length) is measured by a Wheatstone bridge or a potentiometer. With the addition of solute the resistance increases but when the solution becomes saturated it does not increase further. It is possible to pinpoint the solubility from the change in the slope of the resistivity versus composition plot. A typical resistivity versus composition plot is shown in Fig. 9.1. A change in temperature shifts the solubility point, and it is thus possible to measure solubility as a function of temperature.

The technique is simple and involves no sampling procedure. However, the applicability of the technique is limited by the requirement that the resistivity coefficient must be quite high in order to measure low solubilities. On the other hand, the method is unlikely to be influenced by

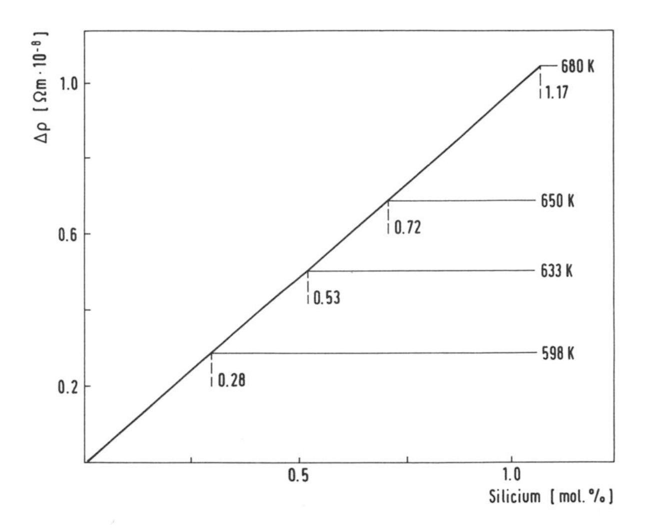

Figure 9.1. Resistivity changes for dissolution and precipitation of Li_4Si.

minor impurities whose resistivity coefficient is less. Table 9.2 lists the resistivity coefficient $d\rho/dc$ of various solute materials in a lithium alloy.

Measurements of lower solubility are carried out by equilibration backed by trace analytical techniques like spectrophotometry, atomic absorption spectrometry, spark source mass spectrometry, radiometry, and activation analysis. The solute is immersed in the liquid alkali metal for sufficiently long time at the temperature of interest so that the liquid metal gets saturated with the solute. The saturated liquid metal is filtered to remove any particulates, and the concentration of the solute is measured using any of the analytical techniques mentioned above. Wherever possible,

Table 9.2. Resistivity Coefficients of Different Solutes in Lithium

Solute	$\frac{d\rho}{dc} \times 10^8$ (Ωm/mol solute)	Temperature (K)
N	7.2	723
H	5.0	723
D	5.0	723
Al	5.8	693
Si	10.4	723
Ge	11.4	723
Na	0.1	723
O	2.1	573

the container for the alkali metal is made of the material whose solubility is to be measured. However, if the solute metal has poor mechanical properties, it is taken as ingots in sodium contained in a suitable material which does not interfere in the measurement. The equilibration is often carried out in a capsule which, in many cases, is a hermetically sealed, evacuated all-metal system. During equilibration the temperature is evenly controlled to avoid particulate formation. Various methods of sampling have been employed. Pressurization of the liquid metal through fine, noninteracting filters and analysis of the collected alloy is a superior method. The sample is analyzed for the solute using trace analytical techniques. Spectrophotometry has been used in the earlier stages of this method, for example, in analyses e.g., Ni, Fe. Of late, atomic absorption spectrometry (AAS) has become the most widely used technique. Whenever the sodium matrix introduces errors, bulk sodium is removed by vacuum distillation. Wherever a suitable radioactive isotope could be found, equilibration has been done with radioactive solute and the concentration of the solute in the sample determined by counting techniques. This latter method has been applied in analyses for Fe, Cr, Cu, Co, Ta, Zr, and other elements in sodium. Activation analysis has been employed in the analysis of some metals, e.g., Co and Nb in sodium. The experimental setup used by Periaswami *et al.*[2] is shown in Fig. 9.2. The equilibration capsule has two compartments, one for equilibrating the solvent with the solute and the other for collecting the sample. The solvent is saturated with the solute in a tantalum container. During equilibration the capsule is held vertically as shown in the figure. Sampling is done by inverting the assembly; the alloy is forced through a stainless steel microfilter by pressurization and collected in a tantalum sample collector. The sample collected is frozen and analyzed for the solute.

Various measures have been taken by different workers to avoid impurities, especially nonmetals that are likely to influence the investigations. Thus oxygen, which influences measurements of solubility of transition metals in liquid sodium, has been removed by gettering. Hot gettering with uranium metal at 973 K has been employed by Stanaway and Thompson.[3] *In situ* gettering with magnesium metal has been used by Periaswami *et al.*[2] The presence of fine particles of the material studied often vitiates the measurement. Filtration using fine sintered disks of a material that does not interfere with the determination is necessary to get good results.

The temperature dependence of solubility can be expressed as

$$\log S = A - \frac{B}{T} \tag{1}$$

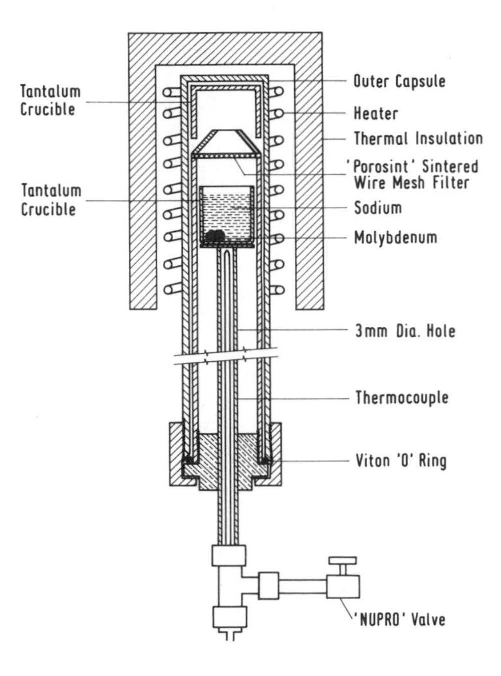

Figure 9.2. Schematic of solubility assembly.

where S is the solubility, T is temperature in Kelvin, and A and B are constants. When the solubility is small, S is given as wppm (parts per million by weight). For higher solubilities, S is given in weight percent. As the second term in the above expression is an enthalpy term, the heat of solution can be obtained from the value of B.

$$\Delta H_{\text{sol}} = \frac{B}{2.303R} \tag{2}$$

In the following sections we discuss the solubility of metals, treating each group of the periodic table separately. Wherever information is available, the solubility data are expressed as a function of temperature. Full phase diagrams of alkali metal–metal systems are outside the scope of this book, but precipitating phases are indicated wherever possible. Data on metal solubilities in higher alkali metals are rather meager. Recommended solubility expressions are given in the Appendix.

9.1. Group IA: Miscibility of Alkali Metals

Liquid alkali metals from Na to Cs are miscible with each other in all proportions. Lithium is the odd man out of the family. Though other alkali metals dissolve in lithium to a limited extent, there is no extensive alloy formation.

Sodium–potassium alloy, called NaK, is the most widely known of alkali metal alloys. As is seen in the phase diagram[4] given in Fig. 9.3, the eutectic composition has 68.1 at% potassium and melts at 260.48 K. The alloy with a potassium concentration of 40 to 90 weight percent is liquid at room temperature. The only compound in the system, Na_2K, melts peritectically, at 280.02 K.

The physical properties of two typical compositions of NaK are given in Table 9.3. It is clear that NaK has good heat transfer characteristics. It is liquid at room temperature and has a wide liquid range. NaK was, in fact, used as coolant in the Dounreay Fast Reactor. It has, however, been superseded by sodium as a fast reactor coolant on account of the latter's superior properties, especially with respect to thermal conductivity, heat capacity, reactivity, and cost. NaK is still used as a coolant in the cold traps of fast reactors.

The sodium–rubidium and sodium–cesium phase diagrams given by Bale[4,5] are reproduced in Figs. 9.4 and 9.5, respectively. The eutectic compositions are at 17.9 and 20.9 at% sodium, respectively. The eutectic melts at 268.65 K in the case of Rb and at 241.32 K in the case of Cs. Only the

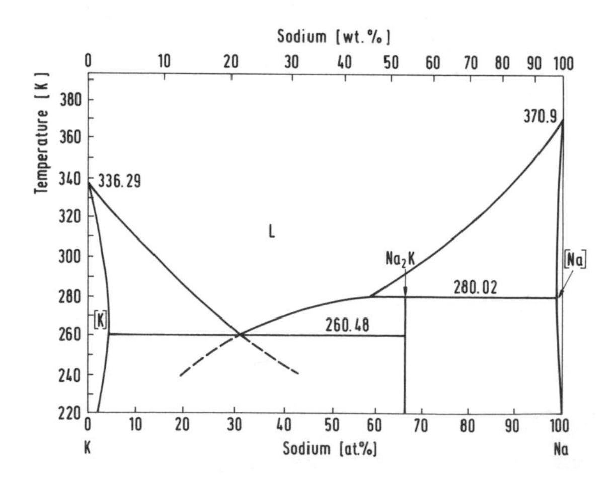

Figure 9.3. Potassium–sodium phase diagram (after ref. 4; reproduced with the permission of the American Society for Metals, Metals Park, Ohio, USA).

Table 9.3. Physical Properties of NaK

	Composition (wt% K)	
	44	78
Melting point, K	292.15	260.48
Boiling point, K	1098.15	1057.15
Density, kg/m^3 (373.15 K)	886	847
Viscosity, mPa · s (523 K)	0.316	0.279
Heat capacity, kJ/kg · K (473 K)	1.096	1.045
Thermal conductivity, W/m · K (473 K)	26.36	25.10

Na–Cs system has a compound (Na_2Cs) and it melts peritectically at 265.1 K.

The potassium–cesium phase diagram shown in Fig. 9.6 is adapted from Hansen and Anderko's compilation.[7] The two elements are fully miscible in both solid and liquid phases. The eutectic appears at 50 at% composition and melts at 235.65 K. The cesium–rubidium diagram is similar.[7]

Ternary alloys of alkali metals are interesting because the eutectics have very low melting points. A ternary eutectic of Na–K–Cs (3% Na–24% K–73% Cs) melts at 197.15 K.[8]

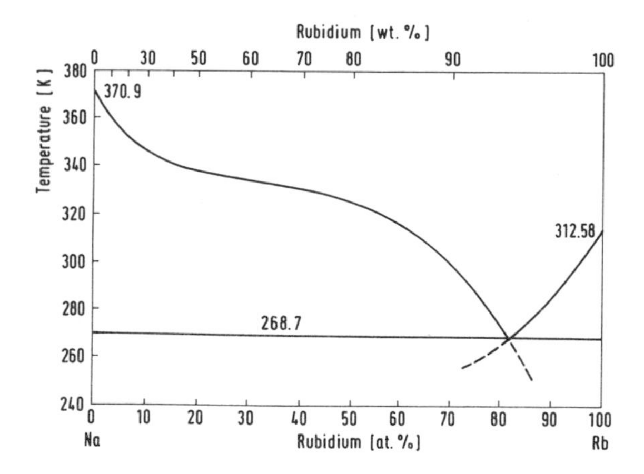

Figure 9.4. Sodium–rubidium phase diagram (after ref. 4; reproduced with the permission of the American Society for Metals, Metals Park, Ohio, USA).

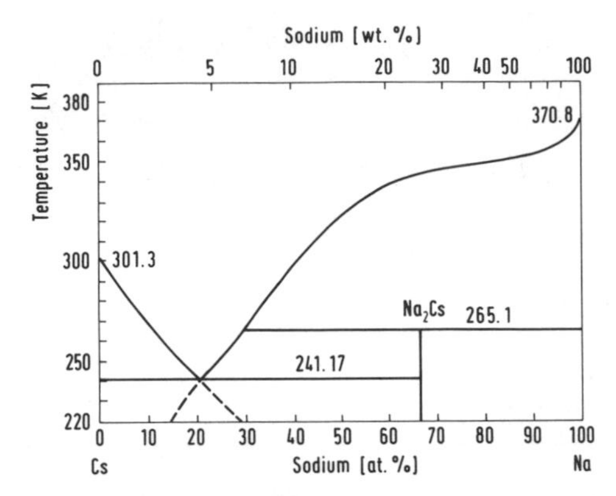

Figure 9.5. Sodium–cesium phase diagram (after ref. 5; reproduced with the permission of the American Society for Metals, Metals Park, Ohio, USA).

9.2. Group IB: Copper, Silver, and Gold

The coinage metals show limited solubility in alkali metals. The solubility increases with increasing size of the solute atoms and decreases with increasing size of the solvent atoms. The tendency for compound formation also shows similar trends. Broad regions of homogeneity occur only in lithium-containing systems. Figure 9.7 shows a comparison of the binary phase diagrams after Kienast *et al.*[9]

Copper has limited solubility in lithium, rising from very low levels near the melting point of the solvent to about 4 at% at 973.13 K.[10] Lithium has substantial solid solubility in copper (~20 at% at the melting

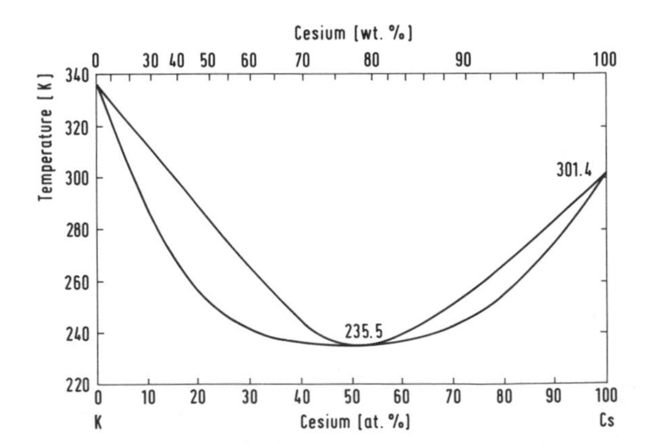

Figure 9.6. Potassium–cesium phase diagram (after ref. 7).

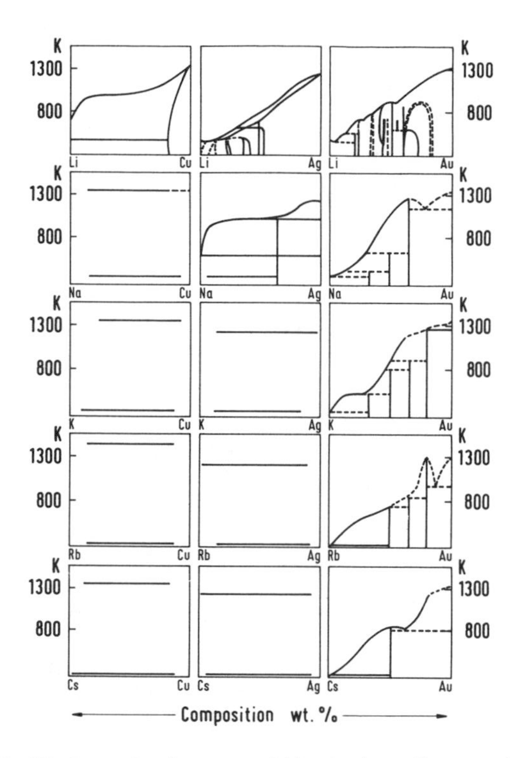

Figure 9.7. Alkali metal–coinage metal binary phase diagrams (after ref. 9).

point of lithium[11]). The solubilities of silver and gold in lithium are much higher. The eutectic compositions are about 10 at% Ag (m.p. 418.65 K) and ~5.1 at% Au (m.p. 424.15 K). These two binary systems are characterized by several compounds, the number being larger in the Li–Au system.

Copper has very little solubility in sodium and in heavier alkali metals. The solubility in sodium has been measured by several groups and these data have been reviewed by Johnson *et al.*[12] and Claar.[13] Claar concluded that fairly good agreement exists among the results of various experimenters in the temperature range 623–773 K. On this basis, he has given a composite solubility curve:

$$\log S\,(\text{wppm Cu}) = 5.45 - \frac{3055}{T(\text{K})} \tag{3}$$

Silver has limited solubility in sodium and much less in higher alkali metals. The Na–Ag phase diagram[9] shows an intermetallic compound,

$NaAg_2$. Lamprecht and Crowther[14] and Weeks[15] have reported the solubility of silver in sodium. These data have been used by Claar[13] to give the following solubility expression:

$$\log S\,(\text{wppm Ag}) = 7.22 - \frac{1479}{T(\text{K})} \tag{4}$$

Gold has greater solubility in all the alkali metals than other coinage metals have. There are several intermetallic compounds in all the binary systems involving gold and alkali metals. The liquidus of the Na–Au system is described by the following expression, given in ref. 12 as valid in the temperature range 373–873 K:

$$S\,(\text{wt\% Au}) = -11 + 0.52T - 6 \times 10^{-4}T^2 \qquad (T \text{ in } ^\circ\text{C}) \tag{5}$$

The solubilities of these elements do not appear to be dependent on the oxygen concentration in sodium.

The cesium–gold binary system shows interesting electrical properties. The binary alloy becomes a semiconductor at the composition of the stoichiometric compound CsAu. The conductivity and thermoelectric power of the alloy drop sharply by several orders of magnitude at the stoichiometric composition.[16] This may be understood in terms of the formation of an ionic compound, Cs^+Au^-, on account of the large electronegativity difference between these elements.

9.3. Group IIA: Alkaline Earth Metals

9.3.1. Beryllium

Very little information is available in the literature on beryllium–alkali metal systems. An approximate solubility of 0.22 wt% (0.17 at%) of Be in lithium at 1273 K has been reported.[17] Though corrosion of beryllium in sodium and NaK has been observed,[18] no solubility data are available. Except for the reported existence of the compound KBe_2,[19] no studies are known of the interaction of Be with the heavier alkali metals.

9.3.2. Magnesium

The diagonal relationship between lithium and magnesium in the periodic table is reflected in the high solubility of magnesium in lithium. The solubility increases continuously with temperature, reaching 30 at% at

673 K. The maximum in the liquidus appears at 867.15 K ($\sim$70 at% Mg). The phase diagram shown in Fig. 9.8a is adapted from ref. 19.

Solubility of magnesium in sodium is less than in lithium. The solubility expression quoted by Johnson *et al.*[12]:

$$S\,(\text{wt\% Mg}) = -0.1414 + 1.248 \times 10^{-3}\,T + 2.08 \times 10^{-6}\,T^2 \quad (T \text{ in } ^\circ\text{C}) \qquad (6)$$

for the temperature range 748–911 K is based on the studies of Mathewson.[20] The phase diagram of the Na–Mg system given in Hansen and Anderko's compilation[7] is also based on the same data. The system needs reinvestigation. The solubility of magnesium is reported to be $\sim$0.8 at% in potassium at 923.15 K, and 1.2 at% in Rb and 1.8 at% in Cs at their respective boiling points.[21]

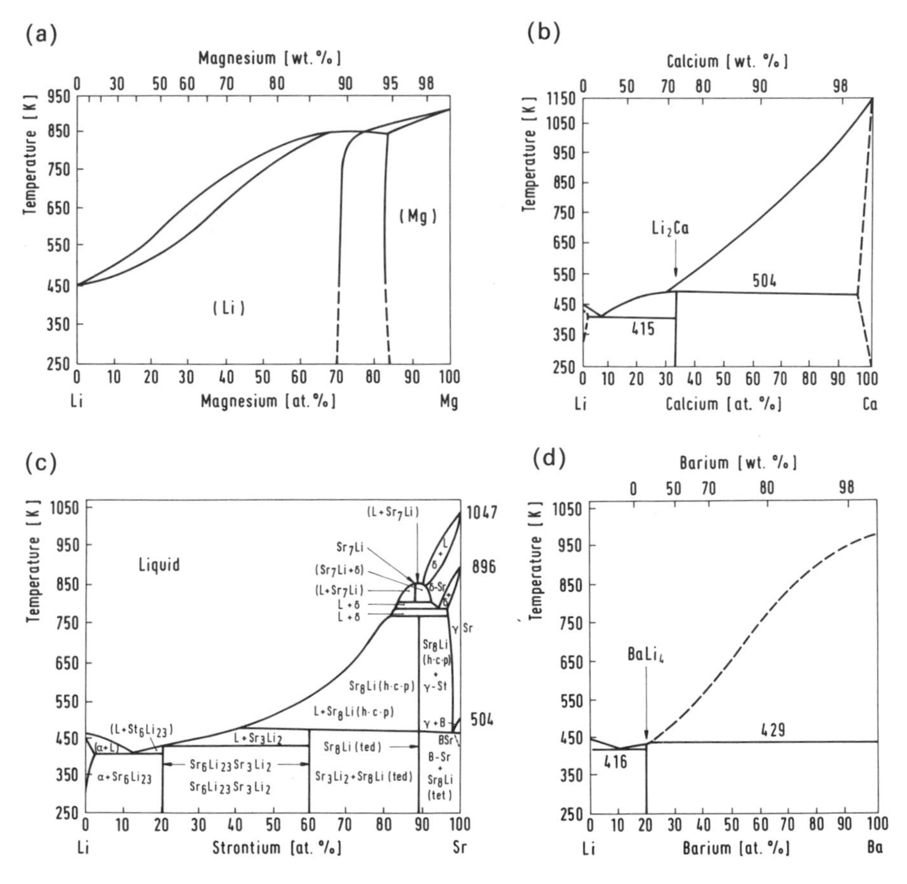

Figure 9.8. (a) Lithium–magnesium phase diagram. (b) Lithium–calcium phase diagram. (c) Lithium–strontium phase diagram. (d) Lithium–barium phase diagram.

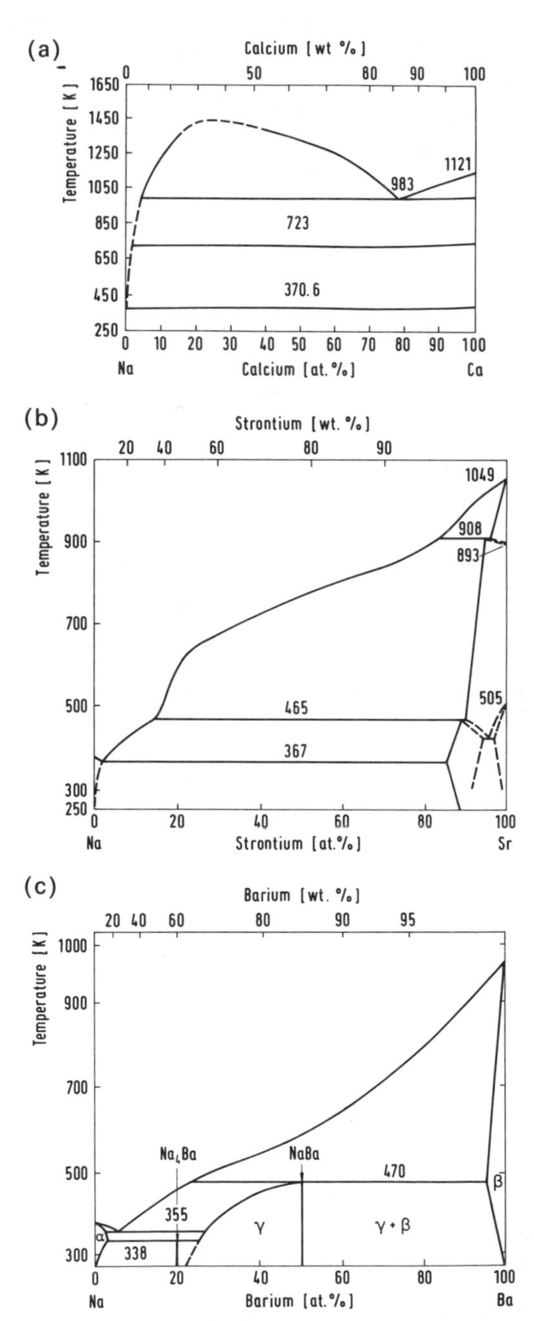

Figure 9.9. Phase diagram of (a) the sodium–calcium system; (b) the sodium–strontium system; (c) the sodium–barium system.

9.3.3. Calcium, Strontium, and Barium

In Figures 9.8b–d and 9.9a–c the binary phase diagrams of lithium and sodium with the alkaline earth metals Ca, Sr, and Ba are compared. The Li–Sr phase diagram is adapted from Wang *et al.*(22) and that for Na–Ba from Kanda *et al.*(23) The rest are based on the compilations of Hansen and Anderko (ref. 7, p. 404), Elliott (ref. 19, pp. 149, 243), and Foust (ref. 12, p. 222). Phase diagrams of heavier alkali metals with alkaline metals are not available.

The solubilities of the alkaline earth metals are the highest in lithium. At 673 K the solubility ranges from ~50 at% for Ca to about 70 at% for Sr. The solubility is much less in sodium, being only 1 at% at 673 K in the case of calcium but increasing with the size of the solute atom to about 60 at% in the case of barium. Intermetallic compounds are observed in most of the binary systems.

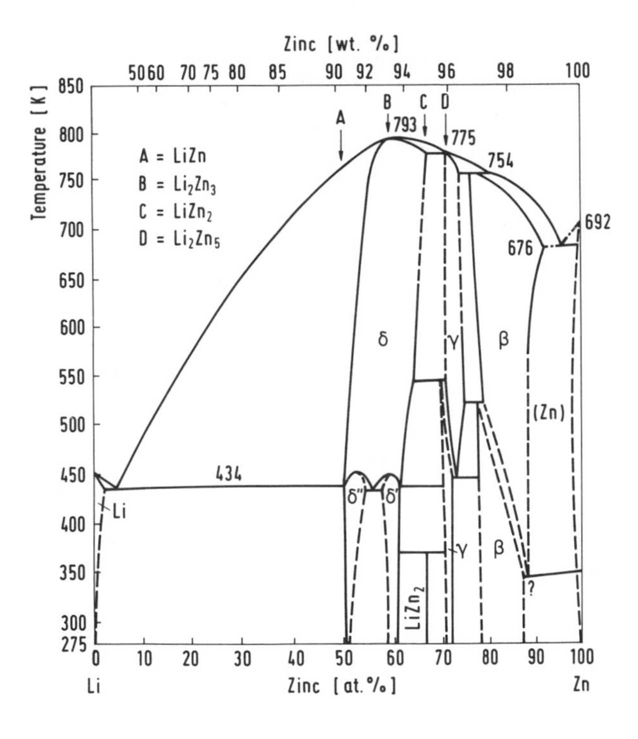

Figure 9.10. Lithium–zinc phase diagram.

(a)

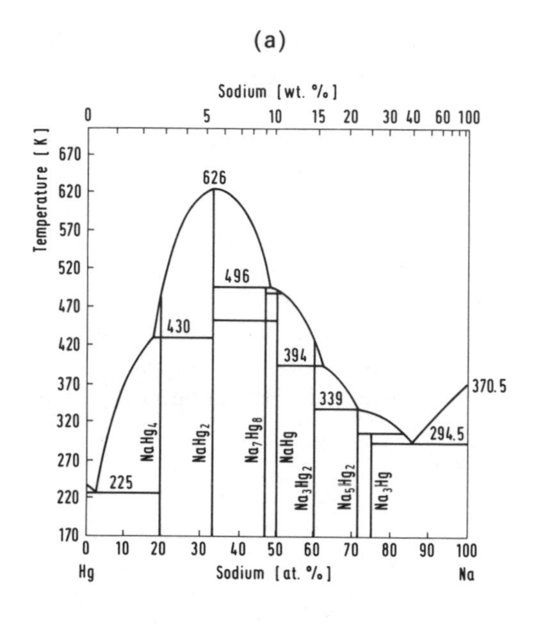

(b)

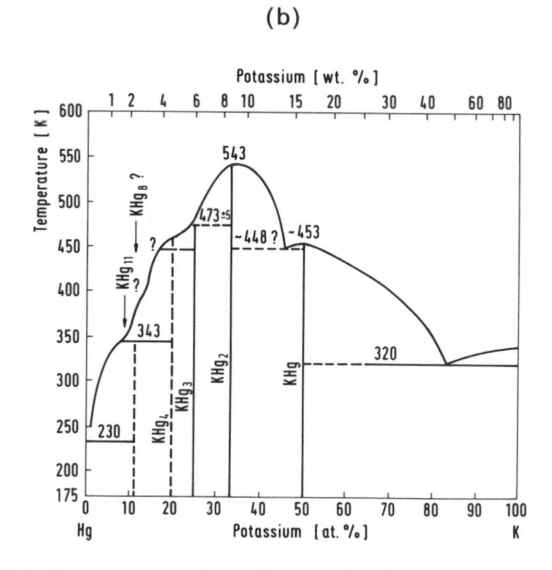

Figure 9.11. (a) Sodium–mercury phase diagram. (b) Potassium–mercury phase diagram.

(c)

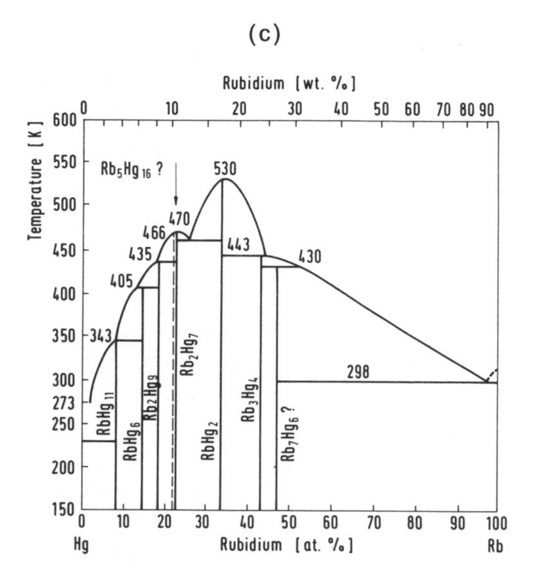

(d)

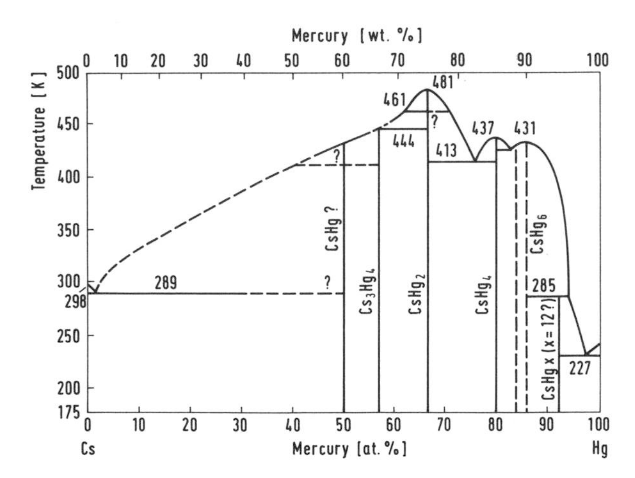

Figure 9.11 *(continued)*. (c) Rubidium–mercury phase diagram. (d) Cesium–mercury phase diagram.

9.4. Group IIB: Zinc, Cadmium, and Mercury

Zinc, cadmium, and mercury exhibit high solubility in liquid alkali metals. The phase diagrams are generally characterized by eutectics at either end with several intermetallic compounds appearing in between. The maxima in the liquids would thus correspond to the melting point of the stablest compound. The melting points of the IIB elements decrease with their atomic number and the maxima in the liquidus curve also decrease in the same order. Figure 9.10 shows a typical binary phase diagram involving lithium (Li–Zn). The phase diagrams involving mercury are displayed in Fig. 9.11.

The lighter group II elements (Zn, Cd) show poor solubility in the heavier alkali metals. Na–Zn[24] and K–Cd[25] are the only systems for which phase diagram information is available. In view of the importance of the Na–Zn system in ^{65}Zn activity buildup in LMFBR primary sodium circuits, the phase diagram is shown in Fig. 9.12. Up to 830 K the solid phase in equilibrium with liquid sodium is $NaZn_{13}$. Above this temperature two liquid phases are present. Lamprecht and Crowther[14] have reported the solubility of zinc in sodium as

$$\log S\ (\text{mole fraction Zn}) = 0.998 - \frac{2652}{T(\text{K})} \qquad (T = 373\text{–}573\ \text{K}) \qquad (7)$$

The solubility of cadmium in sodium is higher. The measurements of Weeks and Davies[25] in the temperature range 373–599 K can be described by the solubility expression:

$$\log S\ (\text{wt\% Cd}) = 3.67 - \frac{1209}{T(\text{K})} \qquad (8)$$

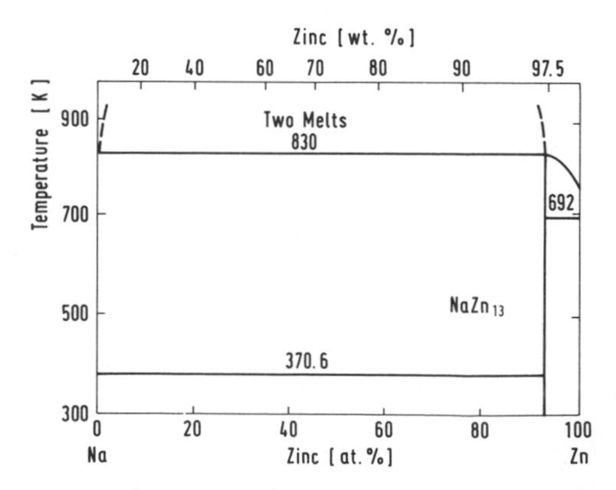

Figure 9.12. Sodium–zinc phase diagram (after ref. 24).

In the cadmium–alkali metal phase diagram the precipitating phase is MCd_{13} where M = Na, K, Rb, and Cs.

9.5. Group IIIA: Boron, Aluminum, Gallium, Indium, and Thallium

Boron does not dissolve in alkali metals to any measurable extent. Binary compounds have been identified in some cases, e.g., NaB_6.[26]

Aluminum shows significant solubility in lithium, but very little in heavier alkali metals. The Al–Li phase diagram[27] given in Fig. 9.13 shows that the melting point of lithium is slightly depressed by the addition of aluminum. The liquidus then rises through the peritectic melting points of the δ and γ phases to the melting point of LiAl, 991 K. At 793 K the solubility is 23 at% Al. The solubility of aluminum in sodium has been represented by Johnson *et al.* (ref. 12, p. 224) as

$$S\,(\text{wppm Al}) = -14.16 + 0.057\,T(\text{K}) \qquad (T = 423\text{–}773\text{ K}) \tag{9}$$

In the Na–Ga (ref. 12, p. 226) and K–Ga (ref. 19, p. 448) systems, the intermetallic compound in equilibrium with the liquid alkali metal phase is M_5Ga_8. The solubility of gallium in sodium in the temperature range 373–573 K is given by the equation

$$\log S\,(\text{at\% Ga}) = 0.867 - \frac{1010}{T(\text{K})} \tag{10}$$

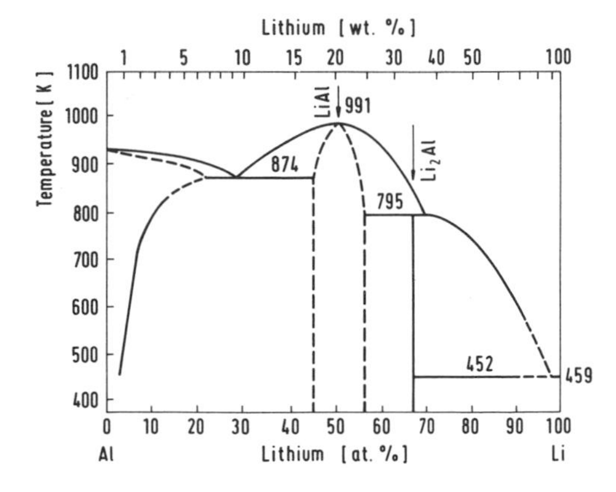

Figure 9.13. Lithium–aluminum phase diagram (after ref. 27).

Gallium solubility in potassium varies from 89 wppm at 573 K to 249 wppm at 823 K.[19]

Indium has good solubility in lithium, reaching 20 at% at 773 K. The solubility of lithium in liquid indium is even greater. The lithium–indium phase diagram given by Hansen and Anderko (ref. 7, p. 848) is based on the observation of the intermetallic compound LiIn. A similar compound has been observed in the Na–In system also, but sufficient information is not available to draw a phase diagram. Lamprecht[28] and Davies[29] have studied the solubility of indium in sodium. Combining their data, Claar[13] has proposed the following solubility expression:

$$\log S\,(\text{wt\% In}) = 4.48 - \frac{1562}{T(\text{K})} \tag{11}$$

which is applicable over the temperature range 373–573 K. The solubility of indium in cesium is reported to be 10 at% at 843 K.[30]

Thallium shows good solubility in lithium (10 at% at 573 K) and lithium dissolves to an even greater extent in liquid thallium. The Li–Tl system is characterized by several compounds ranging from Li_4Tl to LiTl (ref. 7, p. 904). The Na–Tl system is also similar, but the precipitating phase is Na_5Tl (ref. 7, p. 904). The phase diagram based on ref. 7 is shown in Fig. 9.14.

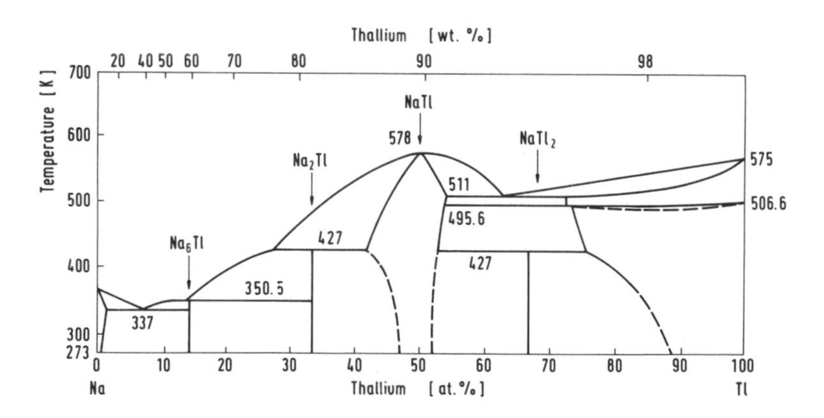

Figure 9.14. Sodium–thallium phase diagram (after ref. 7).

9.6. Group IIIB: Scandium, Yttrium, Lanthanides, and Actinides

The group IIIB elements as well as the lanthanides and actinides have very little solubility in alkali metals. Intermetallic compounds are also not known.

Lamprecht and Crowther's studies[14] suggest that the solubility of cerium in sodium is about 0.06 wppm at 373 to 573 K. Bychkov *et al.*[17] measured the solubility of uranium in lithium. Their data show that the solubility increases from about 50 wppm at 1073 K to 500 wppm at 1273 K.

The determination of the solubilities of uranium and plutonium in sodium is beset with problems, especially due to fine particulates. Caputi and Adamson[31] measured solubilities of urania and plutonia in sodium, making sure that particulates of 5-μm size and above were filtered out. Their results are given in Fig. 9.15. The data can be fitted to the following solubility expressions:

$$\log S\,(\text{wppm, U}) = 4.36 - \frac{6010.7}{T(\text{K})} \tag{12}$$

$$\log S\,(\text{wppm, Pu}) = 8.398 - \frac{10{,}950}{T(\text{K})} \tag{13}$$

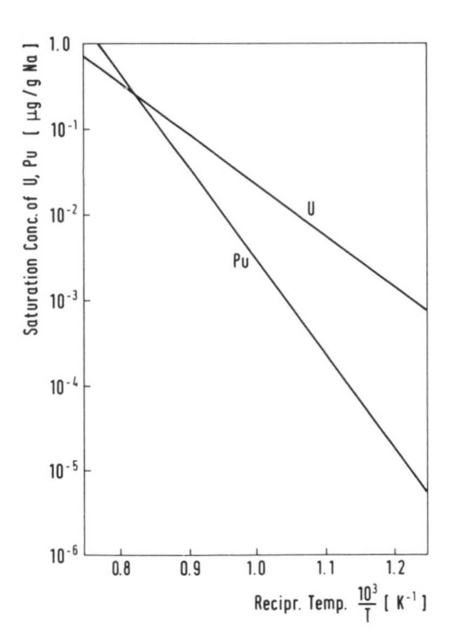

Figure 9.15. Solubilities of urania and plutonia in sodium (after ref. 31).

9.7. Group IVA: Tin and Lead

Tin and lead, the metallic elements of group IVA, show greater solubility in liquid alkali metals than the nonmetallic members of the family. The solubility of lead is higher than that of tin. There are a multiplicity of intermetallic compounds in the binary systems but the precipitating phases are $Li_{22}M_5$ and $Na_{15}M_4$ (M = Sn, Pb).

Hubberstey and coworkers[32,33] have studied the solubility of these elements in lithium and sodium. Their data for tin and lead in sodium are given in Fig. 9.16. The solubility of lead in lithium is described by the equation[34]:

$$\ln S\,(\text{mole fraction Pb}) = 5.717 - \frac{6722}{T(\text{K})} \tag{14}$$

and is valid in the temperature range 550–670 K. The solubilities of lead and tin in sodium fit the following equations:

$$\log S\,(\text{wt\% Pb}) = 6.109 - \frac{2639}{T(\text{K})} \qquad (T = 393\text{–}523\ \text{K}) \tag{15}$$

$$\log S\,(\text{wt\% Sn}) = 5.113 - \frac{2294}{T(\text{K})} \qquad (T = 473\text{–}673\ \text{K}) \tag{16}$$

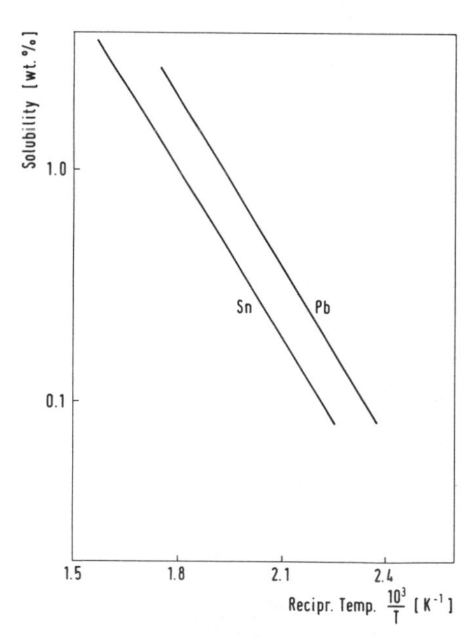

Figure 9.16. Solubilities of tin and lead in sodium (after refs. 32 and 33).

9.8. Group IVB: Titanium, Zirconium, Hafnium

The group IVB elements show very little interaction with alkali metals. The main result of exposing them to liquid alkali metals is oxygen gettering.

Bychkov *et al.*[17] have reported a solubility of 140 wppm for titanium in lithium at 1173 K, whereas Leavenworth and Cleary[35] found it to be less than 10 wppm in the temperature range 973–1173 K. Their data have been fitted by the method of least squares to the following solubility expression[37]

$$\log S\,(\text{wppm Ti}) = 3.351 - \frac{2613.1}{T(\text{K})} \tag{17}$$

This applies to a nitrogen concentration of 55 wppm. Humphreys[38] found the solubility of zirconium in sodium to be 0.09 wppm at 995 K. Bychkov *et al.*[17] found the solubility of zirconium in lithium to vary from 0.00076 at% at 1073 K to 0.023 at% at 1473 K. This agrees with the data of Anderson and Stephen.[36] Pooling these two sets of data we get the following solubility expression:

$$\log S\,(\text{wppm } Z_r) = 9.165 - \frac{8378}{T(\text{K})} \tag{18}$$

Stecura[39] reported the solubilities of titanium and zirconium in potassium as <4 wppm and <10 wppm, respectively, between temperature of 1020 and 1340 K. Eichelberger *et al.*[40] reported a solubility of 6 ± 2 wppm for hafnium in lithium over the temperature range 1273–1673 K. These results show that the solubilities of these elements in alkali metals are very low, but they are not known with a high degree of reliability.

9.9. Group VA: Bismuth

Group VA elements other than bismuth are discussed in Chapter 8. The solubility of bismuth in liquid sodium was measured by Walker and Pratt.[41] These authors have given the following solubility expression for the temperature range 526–836 K

$$\log S\,(\text{at\% Bi}) = 5.0045 - \frac{4189.0}{T(\text{K})} \tag{19}$$

The intermetallic compound in equilibrium with the liquid phase is Na_3Bi (m.p. 1121.25 K). Claar[13] combined Walker and Pratt's data with the results of several American authors and proposed the following two solubility expressions:

$$\log S\,(\text{wt\% Bi}) = 5.67 - \frac{4038}{T(\text{K})} \qquad (T = 563\text{–}923\ \text{K}) \tag{20}$$

$$\log S\,(\text{wt\% Bi}) = 2.15 - \frac{2103}{T(\text{K})} \qquad (T = 398\text{–}563\ \text{K}) \tag{21}$$

9.10. Group VB: Vanadium, Niobium, and Tantalum

The group VB elements have very low solubility in alkali metals. Beskorovainyi *et al.*[42] studied the solubilities of vanadium and niobium in lithium and found them to be less than 3×10^{-3} and 2×10^{-3} wt%, respectively, around 1273 K. The solubility of vanadium in sodium was studied by Babu *et al.*[43] They found the solubility to vary from 0.02 to 0.15 wppm in the temperature range 557–646 K. As at still higher temperatures the data seemed to be affected by the tantalum crucible, they did not propose a solubility equation. Stecura measured the solubility of vanadium in potassium and reported the following solubility expression[39]:

$$\log S\,(\text{wppm V}) = (5.21 \pm 0.26) - \frac{(4526 \pm 298)}{T(\text{K})} \tag{22}$$

This is applicable in the temperature range 1055–1328 K.

Blecherman *et al.*[44] studied the Li–Nb system and reported the following solubility equation applicable in the temperature range 1033–1813 K:

$$\log S\,(\text{at\% Nb}) = -4.77 - \frac{1094}{T(\text{K})} \tag{23}$$

Claar[13] has discussed the few measurements on the solubility of Nb in sodium. The solubility values are scattered in the range of 10 to 100 wppm for temperatures in the range of 873 to 1673 K. This scatter is attributed to the strong oxygen effect on solubility.

The solubility of tantalum in sodium was measured by Grand *et al.*[45] using a radiochemical technique. The reported solubilities range from 2.9 wppm at 798 K to 0.032 wppm at 598 K. Anthrop[37] has suggested that these values could be affected by a systematic error. Ginell and Teitel[46] measured the solubility of tantalum in potassium. The values are scattered, varying from 10 to 150 wppm in the temperature range 1473–1573 K.

9.11. Group VIB: Chromium, Molybdenum, and Tungsten

Chromium and molybdenum being constituents of stainless steels, their solubilities in lithium and sodium have been the subject of several studies. Leavenworth and Cleary[35] measured the solubilities of chromium and molybdenum in lithium. Their data are given in Fig. 9.17. Plekhanov *et al.*[47] also found a similar solubility range for Cr, but a dependence on nitrogen concentration in lithium was observed. More recently, Selle,[48] using Leavenworth and Cleary's data, has given the following solubility equations for Cr in lithium containing different levels of nitrogen, applicable in the temperature range 923–1223 K:

$$\log S\,(\text{atom fraction Cr}) = -2.1364 - \frac{3219.3}{T(\text{K})} \qquad (150\text{ wppm N}) \quad (24)$$

$$\log S\,(\text{atom fraction Cr}) = -1.1138 - \frac{3909.6}{T(\text{K})} \qquad (790\text{ wppm N}) \quad (25)$$

For Mo in lithium, Leavenworth and Cleary's data lead to the following solubility equation:

$$\log S\,(\text{wppm Mo}) = 2.411 - \frac{1509.5}{T(\text{K})} \qquad (26)$$

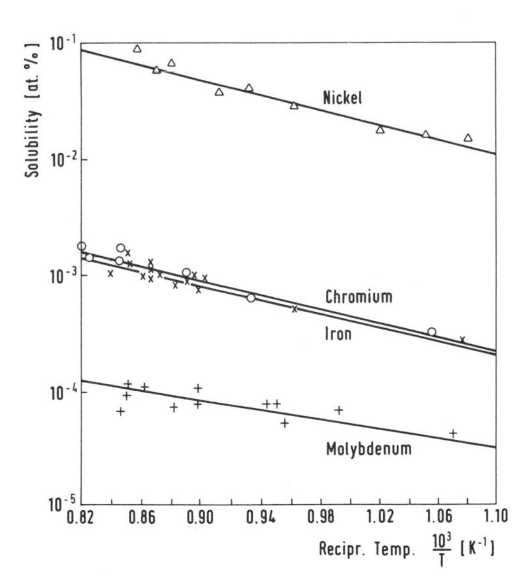

Figure 9.17. Solubilities of stainless steel alloying elements in lithium (based on data given in ref. 35).

Eichelberger *et al.*[40] found the solubility of Mo in lithium to be 4.5 wppm at 1673 K and 1 wppm at 1893 K.

There is considerable scatter in the data reported by Eichelberger and McKisson,[49] Singer *et al.*,[50] and Pellett and Thompson[51] for chromium solubility in sodium. Using the first two sets of data, Singer *et al.* have proposed the following solubility expression[50]:

$$\log S\,(\text{wppm Cr}) = 9.35 - \frac{9010}{T(\text{K})} \qquad (948\text{–}1198\ \text{K}) \tag{27}$$

The solubility of Cr in potassium reported by Ordynskii *et al.*[52] fits the following expression:

$$\log S\,(\text{wt\% Cr}) = 0.158 - \frac{4130}{T(\text{K})} \qquad (1000\text{–}1300\ \text{K}) \tag{28}$$

Molybdenum solubility in sodium was recently investigated by Babu *et al.*[43] in the temperature range 500–720 K. The solubility equation is:

$$\log S\,(\text{wppm Mo}) = 2.738 - \frac{2200}{T(\text{K})} \tag{29}$$

A study of the solubility of molybdenum in potassium by Stecura[39] gave the following expression:

$$\log S\,(\text{wppm Mo}) = 2.21 - \frac{1472}{T(\text{K})} \tag{30}$$

Eichelberger *et al.*[40] reported a solubility of 1 to 2.5 wppm for tungsten in lithium in the temperature range 1473–1873 K. Tungsten solubility in sodium is not known. Stecura[39] has given the following solubility expression for tungsten in potassium in the temperature range 1055–1328 K:

$$\log S\,(\text{wppm W}) = 5.02 - \frac{3851}{T(\text{K})} \tag{31}$$

9.12. Group VIIB: Manganese, Technetium, and Rhenium

Among the group VIIB elements, interest has been limited to manganese. The solubility of manganese in lithium has been studied by

Obinata *et al.*[53] whose data can be fitted to the following solubility expressions:

$$\log S\,(\text{wppm Mn}) = 7.044 - \frac{3199.4}{T(\text{K})} \qquad (T = 773\text{–}973\ \text{K}) \qquad (32)$$

$$\log S\,(\text{wppm Mn}) = 5.646 - \frac{1818}{T(\text{K})} \qquad (T = 973\text{–}1473\ \text{K}) \qquad (33)$$

The solubility of manganese in sodium was reported simultaneously by Periaswami *et al.*[2] and Stanaway and Thompson.[3] Their data are generally in agreement. The solubility equation given by the former authors is reproduced below:

$$\log S\,(\text{wppm Mn}) = 3.6406 - \frac{2601.7}{T(\text{K})} \qquad (T = 550\text{–}811\ \text{K}) \qquad (34)$$

The solubility of rhenium has been reported[40] to be ~ 1 wppm in lithium and ~ 1 wppm in potassium in the temperature range 1473–1873 K.

9.13. Group VIII: Iron, Cobalt, and Nickel

The iron group of elements have very little solubility in alkali metals. The solubility of iron in lithium is strongly influenced by dissolved nitrogen.[17] Plekhanov *et al.*[47] and Anthrop[37] have reported solubility values in the range of 56–80 wppm in lithium containing 90 to 100 wppm nitrogen. Selle[48] gives a solubility equation

$$\log S\,(\text{wppm Fe}) = 4.5539 - \frac{3084.3}{T(\text{K})} \qquad (35)$$

for iron in lithium containing 90 wppm nitrogen (temperature range 933–1193 K).

The solubility of iron in sodium has been measured by many authors, but there is considerable scatter in the data, arising from factors such as the purity of the sodium and the presence of particulates. All the available data were recently reviewed by Awasthi and Borgstedt.[54] Solubility in uranium-gettered sodium was found to vary from 0.3–2.5 wppm in the temperature range 658–997 K and fits the expression

$$\log S\ (\text{wppm Fe}) = 4.720 - \frac{4116}{T(\text{K})} \tag{36}$$

The solubility of iron in potassium has been reported[55] to fit the solubility equation

$$\log S\ (\text{wppm Fe}) = 8.193 - \frac{6166}{T(\text{K})} \tag{37}$$

in the temperature range 900–1200 K.

The solubility of cobalt in lithium is not known. For its solubility in sodium the recent data of Pellett and Thompson[51] may be considered the most reliable. Their solubility equation is:

$$\log S\ (\text{wppm Co}) = 0.010 - \frac{1493}{T(\text{K})} \qquad (T = 673\text{–}973\ \text{K}) \tag{38}$$

The solubility of Co in potassium is less than 5 wppm in the temperature range 900–1200 K.

The solubility of nickel in lithium is higher than that of other alloying elements in stainless steel.[47] Selle,[48] using the data of Leavenworth and Cleary,[35] has reported the following solubility expressions for the temperature range 923–1173 K:

$$\log S\ (\text{wppm Ni}) = 6.662 - \frac{3304.2}{T(\text{K})} \qquad (\text{for 146 wppm N}) \tag{39a}$$

$$= 7.274 - \frac{3002}{T(\text{K})} \qquad (\text{for 220 wppm N}) \tag{39b}$$

Claar[13] has recommended the following solubility expression for nickel in sodium obtained by pooling the data of Singer and Weeks[56] and Eichelberger and McKisson[49]:

$$\log S\ (\text{wppm Ni}) = 2.07 - \frac{1570}{T(\text{K})} \tag{40}$$

This is applicable in the temperature range 573–1173 K.

The solubility of nickel[55] in potassium in the temperature range 941–1328 K has been reported to be

$$\log S\,(\text{wppm Ni}) = 3.89 - \frac{3040}{T(\text{K})} \tag{41}$$

9.14. Ruthenium, Rhodium, Palladium, Osmium, Iridium, and Platinum

The solubilities of these metals in liquid alkali metals are not well established. Several intermetallic compounds of lithium with these noble metals are known. The lithium-rich compounds are of the type LiM (M = Rh, Ir),[57,58] $Li_{15}M$ (M = Pd),[59] and LiM_2 (M = Pt).[60]

Sodium dissolves platinum and palladium in the wt% range. Its melting point seems to remain unaffected by the dissolution of these metals. The compound $NaPt_2$, which can be obtained by the reaction of the component elements, crystallizes in a face-centered cubic lattice of the $MgCu_2$ type.[60] There is some evidence that the heavier alkali metals also dissolve palladium and form intermetallics of the AM_2 type, while the dissolution of other platinum elements in the heavy alkali metals appears to be unlikely on thermodynamic grounds, as has been shown by Niessen *et al.*[61]

Appendix

Table 9.A1. Solubility Expressions for Metals in Lithium

Element	Solubility equation	Temp. range (K)
Pb	$\log S_{wt\%} = 9.192 - 6722/T(K)$	550–670
Ti	$\log S_{wppm} = 3.351 - 2613.1/T(K)$	973–1173
Zr	$\log S_{wppm} = 9.165 - 8378/T(K)$	1100–1500
Nb	$\log S_{wppm} = 1.643 - 1094/T(K)$	1033–1813
Cr[a]	$\log S_{wppm} = 4.738 - 3219/T(K)$	923–1223
Mo	$\log S_{wppm} = 2.411 - 1509.5/T(K)$	933–1183
W	$\log S_{wppm} = 5.02 - 3851/T(K)$	1055–1328
Mn	$\log S_{wppm} = 7.044 - 3199/T(K)$	773–973
	$\log S_{wppm} = 5.646 - 1818/T(K)$	973–1473
Fe[b]	$\log S_{wppm} = 4.5539 - 3084/T(K)$	933–1193
Ni[c]	$\log S_{wppm} = 6.662 - 3304/T(K)$	923–1193

[a] Solubility at 150 wppm N.
[b] Solubility at 90–100 wppm N.
[c] Solubility at 146 wppm N.

Table 9.A2. Solubility Expressions for Metals in Sodium

Element	Solubility equation	Temp. range (K)
Cu	$\log S_{wppm} = 5.45 - 3055/T(K)$	623–773
Ag	$\log S_{wppm} = 7.22 - 1479/T(K)$	377–806
Au	$S_{wt\%} = -11 + 0.52T - 6 \times 10^{-4}T^2$ (T in °C)	373–873
Mg	$S_{wt\%} = -0.1414 + 2.08 \times 10^{-6}T + 1.248 \times 10^{-3}T^2$ (T in °C)	
Zn	$\log S_{wt\%} = 3.452 - 2562/T(\mathrm{K})$	373–573
Cd	$\log S_{wt\%} = 3.67 - 1209/T(\mathrm{K})$	373–600
Al	$S_{wppm} = 1.4 + 0.057 \cdot T$ (°C)	423–773
Ga	$\log S_{wt\%} = 1.349 - 1010/T(\mathrm{K})$	375–573
In	$\log S_{wt\%} = 4.48 - 1552/T(\mathrm{K})$	373–573
U	$\log S_{wppm} = 4.36 - 6010.7/T(\mathrm{K})$	560–970
Pu	$\log S_{wppm} = 8.398 - 10{,}950/T(\mathrm{K})$	560–970
Sn	$\log S_{wt\%} = 5.113 - 2299/T(\mathrm{K})$	473–673
Pb	$\log S_{wt\%} = 6.1097 - 2636/T(\mathrm{K})$	393–523
Bi	$\log S_{wt\%} = 2.15 - 2103/T(\mathrm{K})$	398–563
	$\log S_{wt\%} = 5.67 - 4038/T(\mathrm{K})$	563–923
Cr	$\log S_{wppm} = 9.35 - 9010/T(\mathrm{K})$	948–1198
Mo	$\log S_{wppm} = 2.738 - 2200/T(\mathrm{K})$	500–720
Mn	$\log S_{wppm} = 3.640 - 2601/T(\mathrm{K})$	550–811
Fe	$\log S_{wppm} = 4.720 - 4116/T(\mathrm{K})$	658–973
Co	$\log S_{wppm} = 0.010 - 1493/T(\mathrm{K})$	673–973
Ni	$\log S_{wppm} = 2.07 - 1570/T(\mathrm{K})$	673–973

Table 9.A3. Solubility Expressions for Metals in Potassium

Element	Solubility equation	Temp. range (K)
V	$\log S_{wppm} = (5.21 \pm 0.26) - (4526 \pm 298)/T(\mathrm{K})$	1055–1328
Cr	$\log S_{wppm} = 4.158 - 4130/T(\mathrm{K})$	1000–1300
Mo	$\log S_{wppm} = 2.21 - 1472/T(\mathrm{K})$	1023–1478
Fe	$\log S_{wppm} = 8.193 - 6166/T(\mathrm{K})$	900–1200
Ni	$\log S_{wppm} = 3.84 - 3046/T(\mathrm{K})$	941–1328

References

1. R. J. Pulham, P. Hubberstey, A. E. Thunder, A. Harper, and A. T. Dadd, in: *2nd Intern. Conf. on Liquid Metal Technology in Energy Production* (J. M. Dahlke, Ed.), National Techn. Information Service, Springfield, Va., 1980, (CONF-800401-P2), 18-1.
2. G. Periaswami, V. Ganesan, S. Rajan Babu, and C. K. Mathews, in: *Material Behavior and Physical Chemistry in Liquid Metal Systems* (H. U. Borgstedt, Ed.), Plenum Press, New York, 1982, pp. 411–420.

3. W. P. Stanaway and R. Thompson, in: *Material Behavior and Physical Chemistry in Liquid Metal Systems* (H. U. Borgstedt, Ed.), Plenum Press, New York, 1982, pp. 421–427.
4. C. W. Bale, *Bull. Alloy Phase Diagrams 3*, 313–318 (1982).
5. C. W. Bale, *Bull. Alloy Phase Diagrams 3*, 318–321 (1982).
6. C. W. Bale, *Bull. Alloy Phase Diagrams 3*, 310–313 (1982).
7. M. Hansen and K. Anderko, *Constitution of Binary Alloys*, McGraw-Hill, New York, 1958.
8. J. W. Mausteller, F. Tepper, and S. J. Rodgers, *Alkali Metal Handling and Systems Operating Techniques*, Gordon and Breach, New York, 1967.
9. V. G. Kienast, J. Verma, and M. V. W. Klemm, *Z. Anorg. Allg. Chem. 310*, 143–169 (1961).
10. S. Pastorello, *Gazz. Chim. Ital. 60*, 493, 988 (1930).
11. I. W. Klemm and B. Volavsek, *Z. Anorg. Allg. Chem. 296*, 184–187 (1958).
12. H. E. Johnson, R. L. McKisson, R. L. Eichelberger, and D. C. Gehri, in: *Sodium Na–K Engineering Handbook*, Vol. 1 (O. J. Foust, Ed.), Gordon and Breach, New York, 1972, pp. 169–314.
13. T. D. Claar, *Reactor Technol. 13*, 124–146 (1970).
14. G. L. Lamprecht and P. Crowther, *Trans AIME 242*, 2169–2171 (1968).
15. J. R. Weeks, *Trans. Am. Soc. Met. 62*, 304 (1969).
16. R. W. Schmutzler, H. Hoshno, R. Fischer, and F. Hensel, *Ber. Bunsenges. Phys. Chem. 80*(2), 107–113 (1976).
17. Yu. F. Bychkov, A. N. Rozanov, and V. B. Yakovleva, *Soviet Atomic Energy 7*, 987–992 (1960). (Trans. from *Atomnaya Energiya 7*(6), 531–536.)
18. F. L. Brett and A. Draycott, in: *2nd Intern. Conf. on Peaceful Uses of Atomic Energy*, Vol. 7, IAEA, Vienna, 1958.
19. R. P. Elliott, *Constitution of Binary Alloys*, First supplement, McGraw-Hill, New York, 1965.
20. H. Mathewson, *Z. Anorg. Chem. 48*, 191–200 (1906).
21. W. Klemm and D. Kunze, in: *The Alkali Metals*, Spec. Publ. No. 22, The Chemical Society, London, 1967, pp. 3–22.
22. F. E. Wang, F. A. Kanda, and A. J. King, *J. Phys. Chem. 66*, 2138–2142 (1962).
23. F. A. Kanda, R. M. Stevens, and D. V. Keller, *J. Phys. Chem. 69*, 3867–3872 (1965).
24. E. H. Voice, in: *Liquid Metal Engineering and Technology*, British Nuclear Energy Society, London, 1984, Vol. 1, pp. 259–263.
25. J. R. Weeks and H. A. Davies, in: *The Alkali Metals*, Spec. Publ. No. 22, The Chemical Society, London, 1967, pp. 32–44.
26. P. Hagenmuller and R. Naslain, *C. R. Acad. Sci. 257*, 1294–1296 (1963).
27. A. J. McAlister, *Bull. Alloy Phase Diagrams 3*, 177–183 (1982).
28. G. J. Lamprecht, Ph.D. thesis, University of South Africa, 1966.
29. H. A. Davies, *Trans. AIME 239*, 928–929 (1967).
30. K. A. Chuntonov, L. Z. Melekhov, A. N. Kuzhnetsov, A. N. Orlov, G. G. Ugodniknov, and S. P. Yatseokov, *J. Less-Common Metals 83*, 143–153 (1982).
31. R. W. Caputi and M. G. Adamson, in: *2nd Intern. Conf. on Liquid Metal Technology in Energy Production* (J. M. Dahlke, Ed.), National Techn. Information Service, Springfield, Va., 1980, (CONF-800401-P2), 18-62.
32. P. Hubberstey and R. J. Pulham, *J. Chem. Soc., Dalton Trans.*, 1541–1544 (1974).
33. A. T. Dadd and P. Hubberstey, *J. Chem. Soc., Dalton Trans.*, 2175–2179 (1982).
34. P. Hubberstey, A. T. Dadd, and P. G. Roberts, in: *Material Behavior and Physical Chemistry in Liquid Alkali Metals* (H. U. Borgstedt, Ed.), Plenum Press, New York, 1982, pp. 445–454.

35. H. W. Leavenworth and R. E. Cleary, *Acta Metallurgica 9*, 519–520 (1961).
36. R. C. Anderson and H. R. Stephen, USAEC Report NEPA-1652, August, 1950.
37. D. F. Anthrop, USAEC Report UCRL-50315, 1967.
38. J. R. Humphreys, Jr., quoted in ref. 12.
39. S. Stecura, in: *Corrosion by Liquid Metals* (J. E. Draley and J. R. Weeks, Eds.), Plenum Press, New York, 1970, pp. 601–611.
40. R. L. Eichelberger, R. L. McKisson, and B. G. Johnson, Solubility Studies of Refractory Metals and Alloys in Potassium and Lithium, USAEC Report A1-68-110, 1968.
41. W. A. Walker and J. N. Pratt, *J. Nucl. Mater. 34*, 165–173 (1970).
42. N. M. Beskorovainyi, A. G. Ioltukhovskii, I. E. Lyublinskii, and V. E. Vasilev, *Fiz.-Khim. Mekh. Mater 16*(3), 59–64 (1980).
43. S. R. Babu, G. Periaswami, R. Geetha, T. R. Mahalingam, and C. K. Mathews, in: *Liquid Metal Engineering and Technology*, British Nuclear Energy Society, London, 1984, Vol. 1, pp. 271–275.
44. S. S. Blecherman, G. F. Schenk, and R. L. Cleary, Report CONF-650411, 1965.
45. J. A. Grand, R. A. Baus, A. D. Bogard, D. D. Williams, L. B. Lockard, and R. R. Miller, *J. Phys. Chem. 63*, 1192–1194 (1959).
46. W. S. Ginell and R. J. Teitel, Report CONF-650411, 1965.
47. G. A. Plekhanov, G. P. Fedorstov-Lutikov, and Yu. V. Glushoo, *Soviet Atomic Energy 45*, 818–820 (1979). [Trans. from *Atomnaya Energiya 45*, 143–145 (1978)].
48. J. E. Selle, in: *Intern. Conf. on Liquid Metal Technology in Energy Production* (M. H. Cooper, Ed.), National Techn. Information Service, Springfield, Va., 1976, (CONF-760503), Vol. 2, pp. 453–461.
49. R. L. Eichelberger and R. L. McKisson, USAEC Report ANL-7520, 1968, pp. 319–324.
50. R. M. Singer, A. H. Fleitman, J. R. Weeks, and H. S. Isaacs, in: *Corrosion by Liquid Metals* (J. E. Draley and J. R. Weeks, Eds.), Plenum Press, New York, 1970, pp. 561–576.
51. C. R. Pellett and R. Thompson, in: *Liquid Metal Engineering and Technology*, Vol. 3, British Nuclear Energy Society, London, 1985, pp. 43–48.
52. A. M. Ordynskii, R. G. Popov, G. P. Raikova, N. V. Samsonov, and A. A. Tarbov, *Teplofiz. Vys. Temp. 19*(6), 1192–1197 (1981).
53. I. Obinata, Y. Takeuchi, K. Kurihara, and M. Watanabe, *Metall. 19*, 21–35 (1965).
54. S. P. Awasthi and H. U. Borgstedt, *J. Nucl. Mater. 116*, 103–111 (1983).
55. J. H. Swisher, NASA-TN D-2734, 1965.
56. R. M. Singer and J. R. Weeks, USAEC Report ANL-7520, 1968.
57. S. S. Sidhu, K. D. Anderson, and D. D. Zanberis, *Acta Cryst. 18*, 906–907 (1965).
58. H. C. Darkersloot and J. H. N. Van Vucht, *J. Less-Common Metals 50*, 279–282 (1976).
59. O. Loebich, Jr. and Ch. J. Raub, *J. Less-Common Metals 55*, 67–76 (1977).
60. C. P. Nash, F. M. Boyden, and D. Whittig, *J. Am. Chem. Soc. 82*, 6203–6204 (1960).
61. A. K. Niessen, F. R. de Boer, R. Boom, P. F. de Châtel, W. C. M. Mattens, and A. R. Miedema, *CALPHAD 7*, 51–70 (1983).

10

Corrosion by Liquid Alkali Metals

Corrosion by a liquid metal was the subject of the first publication in which the term "corrodere" (Latin) was used for the reaction of metals with their liquid metal environment. Geber[1] described the phenomenon of the reaction of mercury with gold and silver in his book *De Alchemia* in 1541. Compared to mercury, the liquid alkali metals corrode only a few metallic elements, while they are compatible with transition metals even at elevated temperatures.

The nature of the application of an alkali metal determines the parameters which control the corrosion reactions with solid materials. For instance, the receiver of a sodium-cooled solar energy plant will be heated to about 700°C, a temperature at which corrosion reactions are already considerably fast. However, the flow velocity of the liquid metal in this high-temperature component will be low. Thus, corrosion might be acceptable in spite of the high service temperature, and the component can be used for long periods of time.

In the core of a sodium-cooled fast breeder reactor the temperature will be considerably lower, but the velocity of the coolant has to be high in order to transfer the energy produced within the core to the secondary systems. Therefore, the corrosion of the fuel element clads by the alkali metal is expected to be relatively high. The life of fuel elements is limited by sodium corrosion to just two years at most.

The temperature of sodium–sulfur batteries will not exceed the 300°C level, and the molten metal will be more or less stagnant. Even if the purity of the sodium is poor, corrosion should be low enough to allow long-term operation of such batteries. The storage of rubidium metal in gas cylinders of steel, in which the element is formed by the radioactive decay of ^{85}krypton, a component of the gaseous effluent arising from the reprocessing of fuel elements, will not be affected by liquid metal corrosion, though the storage time is about one hundred years. In the stores the temperature is kept below $\sim 200°C$, and the liquid metal is stagnant under these conditions. Thus, corrosion and mass transfer would be negligible.

10.1. Corrosion Mechanisms

The corrosive action of alkali metals on solid metallic or even ceramic materials may be due to the solubility of metals or nonmetallic compounds in the molten alkali elements. Thus, dissolution of elements as described in Chapter 9 is one of the basic mechanisms of liquid metal corrosion. It has been shown that the dissolution of metals in alkali metals is dependent on temperature. Each individual system of an alkali metal and a solid metallic element behaves as such a liquid solution under certain circumstances. For instance, low impurity levels and high temperatures favor the elemental dissolution of transition metals.

The values for the heat of dissolution of elements in alkali metal melts in the region of metallic dissolution are in the range of $\Delta H_{sol} \sim$ 80,000 J/g-atom. Corrosion reactions based on such solution processes should have activation energies just in this range. The mechanisms of corrosion by liquid alkali metals can thus be deduced from measurements of the temperature dependence of the corrosion reaction rates.

The dissolution of the metals depends on the ratio of the chemical activities of the solute in the solid and in the liquid phase. Dissolution occurs in sodium loops if this ratio acts as a driving force for dissolution. The ratio can be inverted by cooling, however. The opposite process, precipitation of a metallic element out of the alkali melt, takes place in this case. The chemical activity in the solid state is sometimes influenced by the formation of intermetallic compounds. Further, metallic elements can act as traps for dissolved elements, if they have the ability to form such intermetallic compounds.

Exchange reactions of nonmetallic elements between materials and alkali metals may also be the reason for the occurrence of liquid metal corrosion. The chemical activity of oxygen in sodium and the heavier alkali metals influences the corrosion mechanism. The exchange of oxygen often causes the formation of stable oxide layers on the solid alloys. Such layers decrease the corrosion rates as is well known in oxidation reactions in gaseous media. The opposite direction of oxygen exchange has also been observed. In alkali metals of very high purity or very low oxygen activity, oxygen is gettered out of some refractory metals, thus forming layers depleted in oxygen. A special case of oxidation of alloys by alkali metals containing oxygen is internal oxidation, which generates internal layers enriched in oxygen, which precipitates as the most stable oxide.

Oxides formed on the surfaces of materials may be soluble in molten alkali metals. They lose their protective character in that case. The thickness of the oxide layer does not grow with time. The binary metal oxides are sometimes unstable in low-oxygen alkali melts. Oxidation may occur, however, through the formation of ternary oxides of the $A_xB_yO_z$ type. These complex oxides do not possess protective character. Their formation plays an important role in the corrosion mechanisms involving oxygen. The complex oxide studied most extensively is sodium chromite, $NaCrO_2$, the formation of which is observed in corrosion reactions of sodium with austenitic stainless steels.

The direction of the exchange of oxygen between alkali metals and solid materials depends on the thermodynamic stability of the oxides involved in the system. The distribution of oxygen is defined by a distribution coefficient K, which is given by equation (1).

$$K = \frac{a^0_{(s)}}{a^0_{(l)}} = \frac{a^0_{s(s)}}{a^0_{s(l)}} \exp\left[\frac{\Delta G^0_{OX(l)} - \Delta G^0_{OX(s)}}{R \cdot T}\right] \tag{1}$$

In this equation, $a^0_{(s)}$ and $a^0_{(l)}$ are the chemical activities of oxygen in the solid and the liquid phase, respectively, $a^0_{s(s)}$ and $a^0_{s(l)}$ are the chemical activities at saturation, and the ΔG^0_{OX} are the free energies of formation of the oxides stable in contact with the liquid metal. If a complex alkali metal oxide is formed, the activity of oxygen depends on the thermodynamic stability of such a compound, and the ΔG^0_{OX} value of the complex oxide has then to be introduced into equation (1).

The formation of such double oxides in the liquid phase and the precipitation of the compounds formed reduces the chemical activity of the reacting metal in solution. The driving force for dissolution is increased by compound formation, and this increase causes higher dissolution or corrosion rates.

In addition to oxygen, other nonmetallic impurities in alkali metals are also important as reactants in corrosion processes. In lithium corrosion, nitrogen plays a major role by forming stable nitrides, in analogy with the formation of ternary oxides. The complex nitride Li_9CrN_5 is observed in corrosion reactions of lithium with chromium-bearing steels. The reason for the dependence of the corrosion rates on the nitrogen concentration is the formation of such compounds at nitrogen activities sufficient for the formation of complex nitrides. The importance of this reaction is due to the fact that lithium can dissolve considerable amounts of nitrogen, and it is difficult to purify the alkali metal with respect to this element.

The carbon potential of alkali metal melts gives rise to an exchange reaction of this element as a corrosion phenomenon. The decarburization of austenitic stainless steels influences the chemical activity of chromium in the steel. After decarburization of grain boundary areas, the depletion of chromium by alkali metal corrosion will be enhanced. The higher dissolution rate of this element leads to local corrosion effects, such as grain boundary grooving or the formation of cavities along grain boundaries or at edges.

Decarburization of stainless steels by alkali metals is also caused by hydrogen present in the molten metals. Hydrogen reacts with carbon to form methane, which escapes to the gas atmosphere. The carbon activity in the alkali metal is reduced by the methane formation, and thus the driving force for decarburization is increased. This shows that small contaminations of alkali metals with hydrogen are harmful for structural materials. High hydrogen contents, which may be present in the molten metals after reactions with water, give rise to hydrogen embrittlement of steels.

The reactions of element dissolution and nonmetal exchange occur often in parallel. They can both be part of the same corrosion process. For instance, the decreasing of the chemical activity of iron in sodium by the formation of a complex sodium–iron oxide complex enlarges the driving force for the dissolution of iron at high temperatures under conditions which are not influenced by oxygen reactions, as in the high-temperature regions of nonisothermal loops.

The corrosion of alloys by alkali metals is often a combination of several different processes, since elements like nickel or chromium react in individual ways. One of them is based on metallic dissolution and the second on oxidation reactions. Precipitation of corrosion products might be either crystallization of metals or formation of complex oxides as well as other compounds with nonmetals. The deposition of metals may be favored by the formation of intermetallic compounds. Such compounds may be formed by coprecipitation of the dissolved transition elements or by trapping of metals by solid metals or alloys.

10.2. Influence of Parameters

Among the parameters influencing the rates of alkali metal corrosion of solid materials, temperature has the most pronounced effect. Since the elementary processes of corrosion by these liquids are based on the diffusion of metallic or nonmetallic elements, either in the solid materials or in the boundary layers of the liquids, the influence of temperature can easily be understood as a consequence of diffusion. However, the data on the dependence of the corrosion rates on temperature show large scatter, and this fact may be due to the complexity of the corrosion reactions. Thorley and Tyzack[2] have deduced an equation describing the corrosion by sodium of stainless steels in the fast-flowing liquid metal, and this equation has been shown to fit best the authors' results.

$$\log_{10} S = 2.44 + 1.5 \cdot \log_{10}\{\mathrm{O}\} - \frac{18{,}000}{2.3RT} \tag{2}$$

In this equation, S is the loss of material in mils/y (1 mil-0.001 inch), $\{\mathrm{O}\}$ the oxygen concentration in wppm, R the gas constant, and T the absolute temperature. The activation energy of the overall reaction $\Delta H_{act} =$ 18,000 cal/mol (=74,160 J/mol) indicates that the corrosion should be influenced by oxide formation, since the solubility of metals in sodium has a higher activation energy, as shown before.[3]

The results of several studies on the corrosion of steels in sodium containing very low amounts of oxygen indicate that the activation energy is higher under these conditions.[4] This shows that the dissolution of metallic elements may become the rate-controlling step at very low oxygen activities.

In a study on the behavior of pure nickel in a sodium loop over the temperature range 550 to 715°C, the value of $\Delta H_{act} = 28.9$ kcal/mol (=119.07 kJ/mol) was obtained.[5] This value is still higher than the activation energy deduced from measurements of nickel solubility in sodium, thus indicating that in this corrosion reaction the nickel dissolution cannot be the rate-determining step. The diffusion through the laminar boundary layer in the liquid phase might be the rate-controlling part of the process, since the flow velocity in the sodium loop was relatively low.

On the other hand, metals with higher oxygen affinity than the steel elements, for instance, vanadium alloys, exhibit activation energies for weight loss reactions in sodium in the range of 10 to 18 kcal/mol (=41.2 to 74.2 kJ/mol).[6] These low values are due to the fact that oxidation or formation of complex compounds are processes with a lower activation energy.

The temperature dependence of the corrosion of stainless steels by lithium is also strongly influenced by the purity of the liquid metal.[7] Lithium corrosion depends on the nitrogen content of the liquid metal. Ungettered lithium with high nitrogen concentrations leads to values of

$\Delta H_{act} = 20$ kcal/mol ($= 82$ kJ/mol), whereas titanium-gettered lithium leads to a higher activation energy of corrosion of 60 kcal/mol ($= 250$ kJ/mol). The values are not sharply determined, but the corrosion in very pure lithium seems to be comparable to the solution of metals such as nickel in lithium. The formation of the complex nitrides like Li_9CrN_5 seems to become the rate-controlling step in the presence of higher concentrations of nitrogen.

The corrosion rates of steels and nickel alloys in potassium also depend on the test temperature. However, the few measurements of the temperature functions of corrosion rates do not allow a calculation of an activation energy. There is some evidence for a parallel between corrosion by liquid sodium and NaK eutectic and that by potassium. An influence of purity on the slope of the variation in corrosion rate in potassium with temperature was also observed. Most probably, the temperature function of the potassium corrosion of iron- or nickel-based alloys is very similar to that of the sodium corrosion. Refractory metals like vanadium, niobium, or tantalum behave the same in potassium as in liquid sodium. Their corrosion rates are extremely low in very pure potassium. The temperature functions, therefore, could not be estimated with the same precision as in the case of sodium. The corrosion by heavy alkali metals, rubidium and cesium, seems to be dependent on temperature in an analogous way.

Another important parameter in the alkali metal corrosion of materials is the flow velocity of the liquid. It has been shown that in flowing sodium the corrosion rates are proportional to the flow velocity up to values of about 3 m/s, and at higher velocities a plateau of the corrosion rates of stainless steels is reached.[8] It has been claimed that the dependence of corrosion effects on the flow velocities might be due to the thickness of the laminar boundary layers, which decreases with increasing flow velocities (or Reynolds numbers) until a very thin layer remains, which may not be further influenced by the moving liquid. However, it could not be shown if the dependence of the corrosion rates on the flow velocities may also occur for refractory metals or only for steels or nickel alloys. The effect of the flow velocity of lithium on the corrosion rates of stainless steels has also been reported. However, in these studies a plateau value of corrosion rates, which may reach a temperature-independent level, was not detected.

In addition to the flow velocity, the position in a tube through which the liquid metal passes is a parameter influencing corrosion rates. This fact was reported several times as "downstream effect." The corrosion rates decrease with increasing distance from the entrance into the isothermal part of the tube. This decrease of corrosion might be due to chemical changes in the solvents, which can be loaded with the corrosion product or unloaded with the reacting solutes when they flow through the tubes. However, there also is some evidence for hydraulic effects in such tubes. The liquid passing the tube sections may become more and more laminar as it proceeds through the tube. The growth of the laminar boundary layers leads to longer diffusion paths, thus reducing the corrosion rates.

This influence of the position of exposed materials on the corrosion rates has to be taken into account when results of corrosion tests made in different circuits are to be compared. The degree of the downstream effect is dependent on the flow velocity of the liquid alkali metal, as is shown for liquid sodium in Fig. 10.1.[9] The position of the specimens can be normalized by using the "downstream length" L/D, with L as the distance from the entrance of the isothermal tube and D as the hydraulic diameter of this tube. At high L/D values, the corrosion can be considerably reduced by this effect. The downstream effect has also to be introduced into

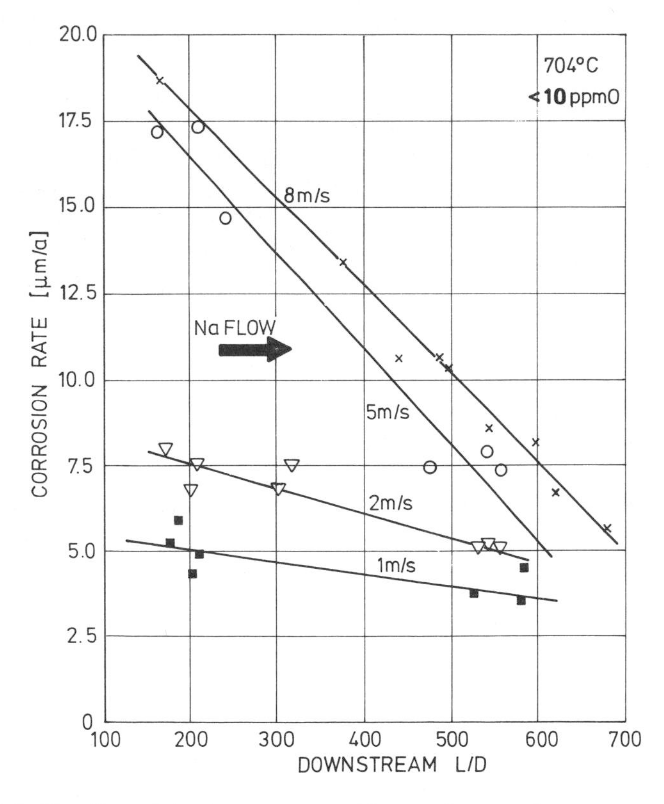

Figure 10.1. The effect of the downstream position on the corrosion rate over a range of flow velocities in sodium at 704°C with less than 10 wppm of oxygen (after ref. 9; reproduced with the permission of Argonne National Laboratory).

corrosion models describing the corrosion and mass transfer of a system of alkali metal circuits.

The corrosion of steels and even austenitic steels by alkali metals is influenced by their impurity contents. In the heavier elements of this group, i.e., sodium, potassium, rubidium, and cesium, it is the oxygen content which increases the corrosion rates with increasing concentration. It has been shown that in case of corrosion of steels by fast-flowing sodium, the influence of the oxygen concentration is stronger than linear, as can be seen from equation (2). It seems that there exists a lower threshold below which the oxygen effect is very small. This threshold might correspond to a certain oxygen activity above which the complex sodium chromite, $NaCrO_2$, becomes thermodynamically stable. However, the formation of other oxides can also be the reason for an oxygen effect on the corrosion rates. The free energy of formation of sodium chromite is a function of the temperature.[10] Therefore, the threshold concentrations of oxygen in sodium as well are temperature dependent.

The rates of corrosion of vanadium alloys in liquid sodium are strongly influenced by the oxygen concentration. While at very low oxygen contents—at 1 wppm and below—the corrosion reaction is an internal oxidation of the alloys, higher oxygen contents cause the formation of a nonadherent oxide layer. In this case severe weight losses occur.

The corrosion of nickel-based alloys by molten alkali metals seems to be independent of the oxygen potentials. This might be due to the fact that complex alkali metal–nickel oxides are not stable under the conditions considered here.

The formation of complex chromium and iron oxides at particular oxygen activities has not only been shown in sodium, but also in heavier alkali metals. For instance, in the stainless steel–rubidium–oxygen system several complex oxides such as Rb_3CrO_4, Rb_4CrO_4, and $Rb_6Fe_2O_6$ have been identified as corrosion products, and their free energies of formation were measured during the corrosion tests by means of EMF methods.[11]

An attempt was made to normalize the corrosion rates of the austenitic steel AISI 316 with respect to the oxygen potential in the liquid alkali metal. The measured corrosion rates were divided by the oxygen concentration.[12] This model, however, did not lead to a good agreement between the data obtained in the different laboratories.

In the lithium corrosion of these materials, nitrogen plays the same role as oxygen in sodium or the heavier alkali metals. Though a quantitative relationship between corrosion rates and the nitrogen concentrations has not yet been established, the corrosion rates are indeed influenced by the nitrogen content of lithium, as is shown in Fig. 10.2. In nitrogen-containing lithium, the formation of lithium chromium nitride,

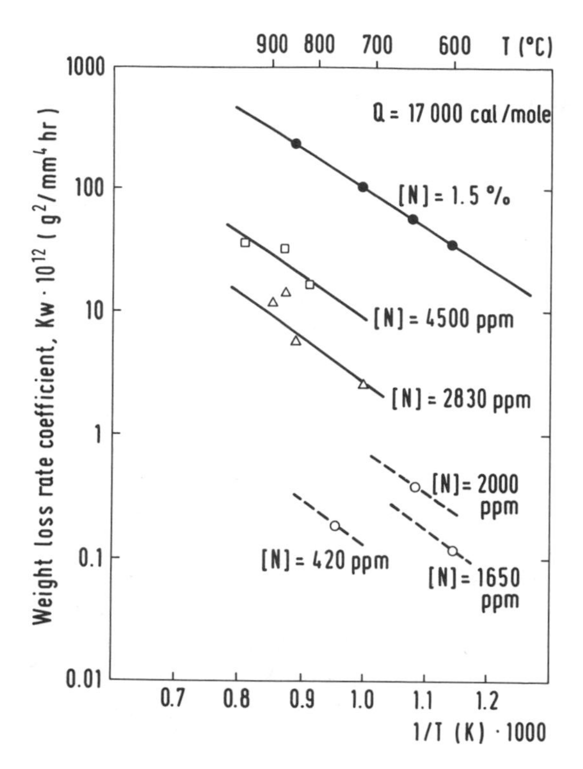

Figure 10.2. The influence of the nitrogen content in lithium on the corrosion rates of stainless steels.

Li_9CrN_5,[13] is the main corrosion reaction. This compound is preferentially formed at the grain boundaries of chromium–nickel steels. Its formation leads to a leaching of chromium, which decreases the strength of the material.

The carbon potential in the alkali metals causes carbon exchange between materials and alkali metals. Since the carbon solubility in the molten metals strongly depends on the temperature, the chemical activity of this element decreases with rising temperature. Thus, alkali metals generally become decarburizing when the temperature of the system is raised. The carbon exchange between sodium and austenitic steels has been extensively studied because of its importance for the structural material of fast breeder reactors.[14] Even models describing the direction and kinetics of the carbon exchange have been developed. Thermodynamics and diffusion processes control the carbon exchange. Solid carbides such as $Cr_{23}C_6$ are formed and get precipitated in the surface layers of austenitic steels. The concentration gradients of carbon correspond to the diffusion of the element in grain boundaries of the steel in the layer below the precipitates. In case of high-temperature corrosion of steels, the formation of ferrite layers can be superimposed on carburization. Carbon concentration profiles indicating carburization have been observed in tests with several austenitic steels. Figure 10.3 shows a result obtained in corrosion experiments with such an austenitic steel, X6 CrNi 18 9. The steel has been exposed in a loop in which sodium circulated at 700 and 600°C.[15] The examination showed that the fuel element cladding tubes of 0.4-mm wall thickness were nearly completely carburized by the sodium. The effect at 700°C caused the depletion of carbon in the first 30 μm adjacent to the surface. This layer is also depleted in chromium and austenite-stabilizing metals. The structure is converted to the ferritic one. The resulting austenitic matrix below the ferrite layer is enriched in carbon. The maximum of the carbon concentration occurs at about a 50-μm distance from the surface. The gradient of the carbon concentration corresponds to the diffusion behavior of carbon in austenitic stainless steel. The maximum of the carbon content is closer to the surface after an exposure at 600°C. The ferrite layer is very thin in this case. The higher value of the concentration at the maximum is due to the higher carburization potential of sodium at lower temperature. High-temperature sodium tends to decarburize unstabilized austenitic steels.

The carbon exchange in systems of solid materials and liquid lithium is very similar to that described for sodium.[14] The carbon exchange reactions in the heavy alkali metals have not yet been studied. Carbon exchange also plays an important role in the corrosion of vanadium alloys by liquid lithium. Carburization layers in vanadium alloys form concentration profiles typical of diffusion processes.

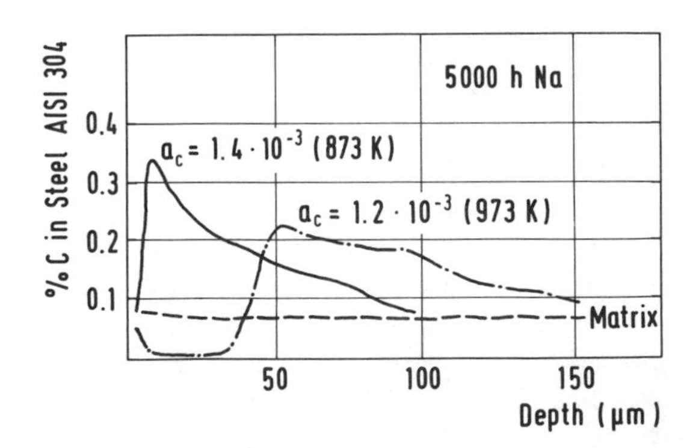

Figure 10.3. Carburization of the austenitic steel X6 CrNi 18 9 (AISI 304) after 5000 hours of exposure to flowing sodium at 600 and 700°C.

10.3. Impact of Alkali Metal Corrosion on the Mechanical Properties of Materials

The exchange of minor elements like carbon or nitrogen between alkali metals and metallic materials causes changes of their creep-rupture strengths and ductilities. The high-temperature strength of austenitic steels depends on the sum of the concentrations of both minor elements. Creep-rupture tests under decarburizing conditions in sodium at 700°C have resulted in considerably reduced time-to-rupture of the specimens of steel type AISI 316, as is shown in Fig. 10.4.[16]

Even at temperatures of about 550°C sodium reduces the time-to-rupture and the rupture-strain of creep-stressed specimens of austenitic steels such as AISI 304. The effect of sodium influences the last period of the creep-rupture life, the range of tertiary creep. The effect has, therefore, been called "tertiary creep embrittlement."[17] The diagrams of Fig. 10.5 illustrate this effect. In (a), creep curves of such specimens tested at 550°C in air and in sodium are compared.[18] The shorter period of tertiary creep in the sodium test is the reason for a reduced lifetime as well as a lower value of strain-to-rupture. Part (b) of the figure shows how the combined effect of sodium and stress (or strain) gives rise to the formation of surface cracks all over the deformed part of the specimens. These surface cracks are

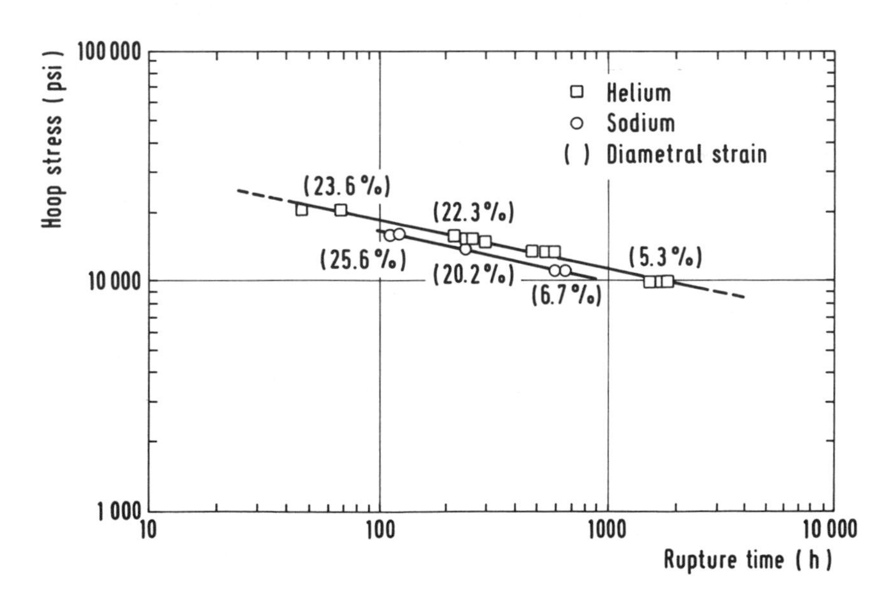

Figure 10.4. Influence of gettered sodium on the creep-rupture life of specimens of steel AISI 316 (after ref. 16).

(a)

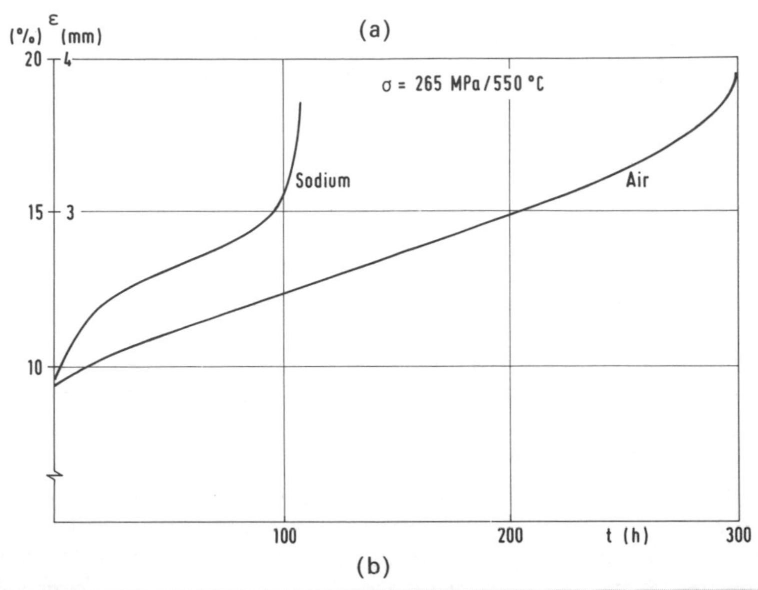

(b)

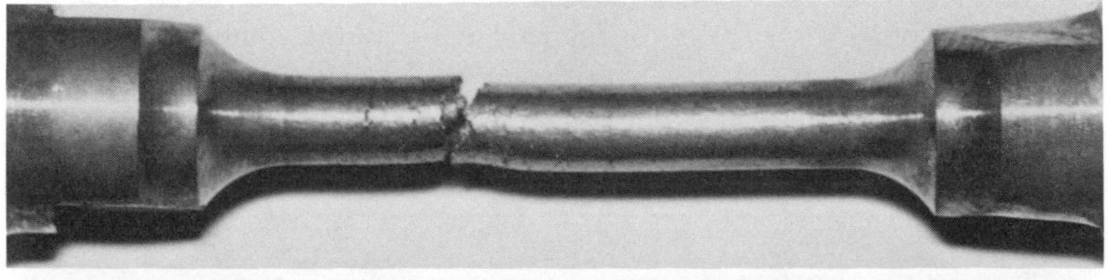

(c)

Figure 10.5. "Tertiary creep embrittlement" of steel AISI 304 by flowing sodium. (a) Creep curves in air and in sodium. (b) Formation of surface cracks in tests on the sodium loop. (c) Crack after failure of the specimen.

intergranular. Their formation is the first step of the process leading to the failure of the steel, as is shown in part (c) of the figure. Scanning microscopic examinations of such specimens indicate that the action of sodium seems to form chains of small holes along the surface grain boundaries before the development of surface cracks. The chains of microscopic holes grow to form intergranular cracks as the material is strained further. It is not yet clear if the formation of such germs of grain boundary cracks might be due to the dissolution of small particles originally precipitated in the grain boundaries. In some of the in-sodium creep-rupture tests the effect of sodium was a pronounced one. The examination of the specimens indicated that there was a decarburization of the specimens. The determinations of the final carbon content were in agreement with carbon activity measurements in sodium. Such a decarburization can be the reason for the dissolution of precipitated carbides and for reduced strength of the steel as well.

The preferential loss of chromium is also observed with the same liquid metal parameters. Local decreases of the chromium concentrations cause local increases of the carbon activity, thus forcing the dissolution of carbon by the alkali metal.

The steel AISI 316 does not seem to be sensitive to this kind of sodium effect in the same way. The higher stability of the austenite structure of this steel might be the reason for the higher resistance against the effect of sodium.

Such effects on the mechanical properties of materials have not yet been observed in systems with other alkali metals. However, one has to take into account the fact that these metals can also act as decarburizing media, thus causing similar effects on the mechanical properties.

Refractory alloys, which also suffer exchange of interstitial elements with alkali metals, are influenced in their mechanical behavior as well. The formation of brittle surface layers is caused by nitriding and carburizing. These effects reduce the creep rate and influence the time- and elongation-to-rupture as well. The opposite direction of minor element exchange lowers the strength and raises the ductility of such alloys.

10.4. Corrosion Phenomena

The comparison of the corrosion rates of pure iron and iron-based alloys like the austenitic stainless steels indicates that the temperature dependence of the corrosion rates in liquid sodium is the same for the whole group of these materials. The corrosion phenomena, however, are different, since leaching of alloying elements by the liquid metal does not occur in the reactions of iron as it does in sodium–steel systems. The corrosion of iron is, therefore, a simple solid phase–liquid phase interface process. Characteristic grain boundary grooves are formed on the surfaces of iron sheets exposed in a sodium loop at a temperature of 600°C. The surface grains exhibit dissolution lines of a terrace-like appearance.

Ferritic steels such as the 2.25 Cr 1 Mo (with or without Nb content) exposed to flowing sodium at temperatures of up to 650°C form a refined grain structure. This layer, however, is a very thin one, and no changes in the chemical composition are detected. The reason for the formation of these small grains may be the loss of carbon out of the surface zone of the material due to the decarburizing action of the alkali metal.

The corrosion phenomena are sensitive to changes in the temperature and oxygen concentration of the alkali metals. As shown above, the temperature dependence of the corrosion rates is very similar for all types of iron-based alloys. The corrosion rates are also dependent on the oxygen concentration in the alkali metals. The dependency has the form $(c_O)^n$ with n close to 2. The effect of decarburization depends on the original microstructure of the 2.25 Cr 1 Mo steel. The precipitation of carbides of the $M_{23}C_6$-, M_6C-, and M_7C_3-type by a normalizing tempering procedure at 650°C makes the steel less sensitive to carbon losses when exposed to alkali metals.[19]

Even ferritic steels with higher contents of alloying elements do not form layers of different chemical composition due to the leaching of elements by the alkali metals. The ferritic–martensitic steels with 12 wt% chromium develop a surface appearance which looks etched. The martensitic needles withstand the attack of the alkali metal better than the remaining austenite around them, and thus a relief of needles is formed.

High oxygen concentrations in the alkali metals cause formation of alkali metal chromite layers. A concentration of 25 wppm oxygen in liquid sodium is high enough to form such layers on the surfaces of 9 wt% chromium steels. Such a layer is very stable against removal by the flowing alkali melt. There is, however, some interference of this surface oxide with the steel matrix. Internal oxidation occurs in the zone adjacent to the surface oxide due to this reaction. Microprobe analyses of the precipitated oxides show that sodium chromite, $NaCrO_2$, is also among the products of internal oxidation.

Austenitic steels with 16 to 20 wt% chromium and 8 to 16 wt% nickel are leached out by flowing alkali metals. The austenitic structure becomes unstable due to losses of chromium, nickel, and manganese. The 18 Cr 9 Ni steel forms more or less thick layers of material depleted in this way and transformed into the ferritic structure. The austenitic structure of this steel is relatively unstable, and even small losses of nickel and manganese are sufficient to cause a change of the structure.

Austenitic steels containing higher nickel concentrations have a more stable austenitic structure. The formation of only very thin ferritic layers and, additionally, layers in which the grain boundaries have changed to the ferritic structure is caused by the leaching out of the austenite-stabilizing elements. Figure 10.6 gives an illustration of the formation of grain boundary ferrite in a stabilized 16 Cr 16 Ni steel. Interference methods have been applied to make the ferrite visible in the metallographic cross sections.

Concentration profiles of alloying elements which have been subjected to the element-leaching process can be measured by means of glow discharge optical spectrometry.[20] Figure 10.7 shows results of such profile analyses made on the surface zone of the steel X8 CrNiMoVNb 16 13 after an exposure in sodium at 700°C for 5,000 hours. There are considerable losses of chromium, nickel, and manganese in the outer layer (0.005 mm from the surface). This layer is the ferritic surface zone shown by

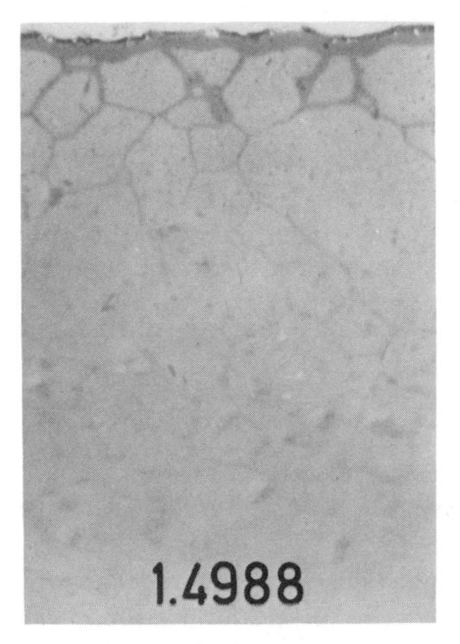

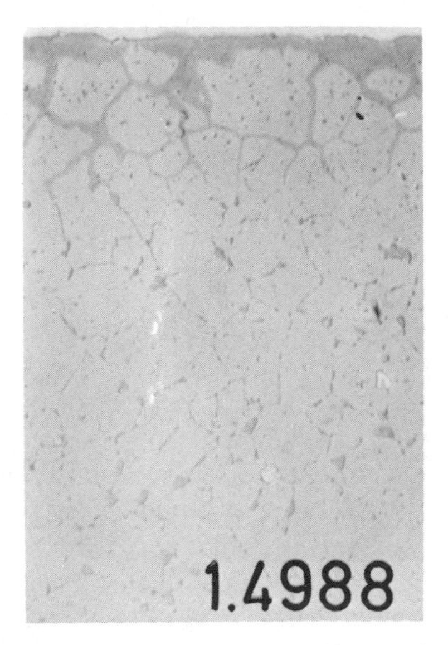

Figure 10.6. Cross-sectional views of specimens of the austenitic steel X8 CrNiMoVNb 16 13 after exposure to flowing sodium (700°C), showing surface ferrite layer developed by ZnSe vapor deposition.

metallographic techniques. The following layer shows a flat gradient of the concentrations of these elements, corresponding to the region with austenitic grains and ferritic grain boundaries. There is nearly no carbon present in the ferritic layer, the measured concentrations being close to the blank of the method. The layer of grain boundary ferrite contains more carbon than the matrix, since the austenite grains in this layer are carburized, as shown earlier. The leaching of substitutional elements like nickel or manganese has not only caused a change of the structure, but also the formation of cavities, mostly at triple grain boundaries. Such cavities extend into the second corrosion layer of steel specimens.

The driving force for the selective leaching of nickel is the low value of its chemical activity in the bulk flow of sodium. The chemical activity at the steel surface becomes equal to the value in the boundary layer of the sodium at the interface. There is a steep activity gradient in the boundary layer. The transfer of nickel through this layer is based on diffusion processes. These are fast since the thickness of the layer is very small, particularly at high flow velocities of the liquid metal. The concentration profile of nickel as indicated by the profile of its chemical activity is also due to diffusion processes (see Fig. 10.8). This figure indicates that the diffusion in the steel AISI 304 is partly grain boundary diffusion. Therefore, the evaluation of diffusion coefficients based on the measured concentration

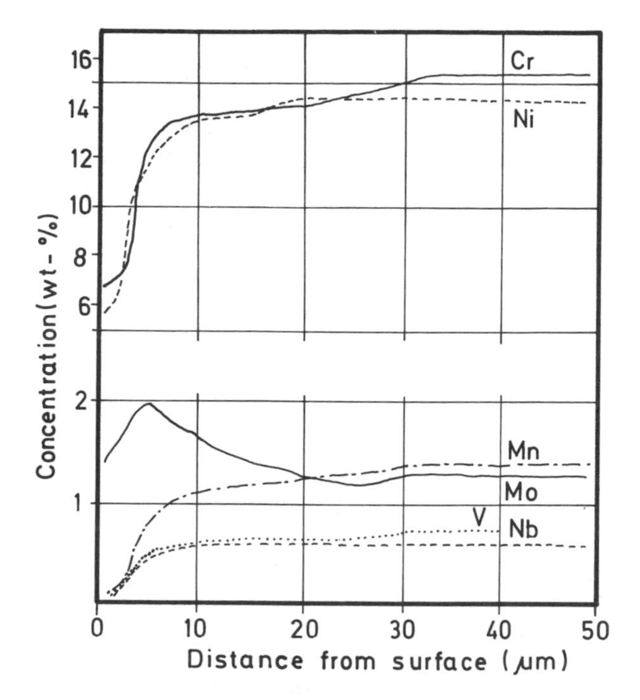

Figure 10.7. Profile analyses of the corrosion zone of the 5,000-h specimen shown in Fig. 10.6.

profiles results in an effective diffusion coefficient which contains contributions from both grain boundary and volume diffusion.

Liquid lithium causes the same type of corrosion effects on austenitic steels as sodium does. The effects are, however, much stronger in lithium, as is shown in Fig. 10.9a. The ferritic surface layer formed by lithium corrosion is the remaining fragment of the original surface. It is very porous due to cavities larger in dimensions and number than in liquid sodium. Figure 10.9b indicates that in the steel X10 CrNiMoTi 18 10, nickel and manganese are drastically leached out. The remaining ferritic material is enriched in chromium, molybdenum, and iron.

Alloys richer in nickel than the austenitic steels, for instance, Incology 800®, are attacked in the same way as steels, with sodium leaching out the alloying elements nickel, chromium, and manganese. This does not cause, however, the loss of the austenitic structure, which is very stable in this alloy. Losses of substitutional components cause the growth of cavities as in the case of steels. It has not yet been ascertained whether this formation of cavities is caused only by the loss of substitutional

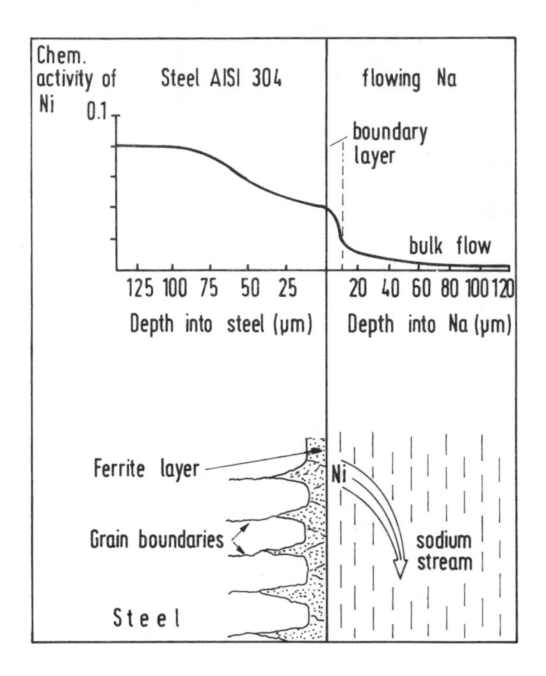

Figure 10.8. The activity difference between solutions of nickel in stainless steel and in sodium as driving force of selective leaching of nickel.

elements or also by dissolution of the precipitated γ'-phase. Nickel-based alloys also do not form layers of changed structures. Their corrosion reactions are much faster, and weight losses correspond to losses of thickness of the specimens.

Alloys based on group IV and V metals such as titanium, zirconium, vanadium, niobium, or tantalum show different corrosion phenomena after an exposure to liquid alkali metals. They do not suffer a leaching of substitutional elements. The main corrosion reaction is the exchange of nonmetallic elements between the alkali metal melts and the alloys. As far as the exchange of oxygen is concerned, lithium is the only alkali metal which is able to reduce most of the alloying elements of this group. If vanadium or niobium contains oxygen in the form of oxides precipitated at grain boundaries, these phases can be dissolved by lithium. Cavities are the results of the dissolution of precipitates. As an example, Fig. 10.10 shows the corrosion effect of lithium on unalloyed vanadium. The observed cavities are caused by the loss of oxygen which was present in the metal.

The oxides of the heavier alkali metals are less stable than the oxides of the group IV and V elements. Therefore, oxygen is moved in the opposite direction. Several corrosion effects may occur by oxidation of the refractory alloys. Zirconium exposed to liquid sodium forms a surface oxide stable against removal by the molten metal. The oxide has protective character, causing a parabolic corrosion rate law. The oxide layer is able to react with the matrix at higher temperatures, thus also forming a zone which contains dissolved oxygen. Precipitation of oxides may occur in this zone of the alloy, if the saturation of the metal with respect to the oxide solubility is exceeded.

Titanium does not form surface oxide layers under the same conditions. Most probably, the oxide has a solubility in the liquid phase. This solubility enables the liquid metal to dissolve the oxide as soon as it is formed by oxygen exchange. Some oxygen uptake by the matrix can also be observed. The corrosion kinetics are not governed by the diffusion of ions through an oxide layer, and thus they follow a linear rate expression.

Vanadium alloys show different phenomena of alkali metal corrosion depending on the oxygen potential of the alkali melt. Internal oxidation is the main corrosion reaction at low oxygen activities. The growth of the internal oxidation layers is governed by a parabolic rate expression. In case of higher oxygen potentials, the internal oxidation is superimposed by surface oxidation and dissolution of the oxide. Therefore, the reaction product does not show protective character. The weight losses measured are relatively high. The zones of internal oxidation and oxygen dissolution are of high hardness and low ductility, and in severe cases they are brittle.

Nitrogen dissolved in alkali metals causes similar types of corrosion phenomena in group IV and V alloys, mainly the formation of internal zones of nitrogen uptake and nitride precipitation. The outer layers of the alloys exchange nitrogen until the equilibrium nitrogen distribution is reached, as described in Chapter 6. In the lithium corrosion of vanadium alloys, the nitrogen exchange can reach a considerable degree. Nitrides are precipitated and cause the hardening of the penetrated zones, as is illustrated in Fig. 10.11.

Additionally, the internal zone of nitride precipitation reacts with the matrix of the alloys, and nitrogen is exchanged between the two regions of the alloys. In these zones, the nitrogen concentrations remain below the saturation limits. Concentration profiles indicate that the migration of nitrogen into the matrix is governed by nitrogen diffusion.

The group VI metal molybdenum has an excellent corrosion resistance against liquid alkali metals. Though the formation of sodium molybdate, Na_2MoO_4, is postulated, it does not cause corrosion rates above 0.035 mm/a at temperatures from 710 to 760°C in flowing sodium. The molybdenum-based alloy TZM (Mo–0.5Ti–0.1Zr) suffers very little intergranular attack by sodium vapor at temperatures as high as 1300°C.

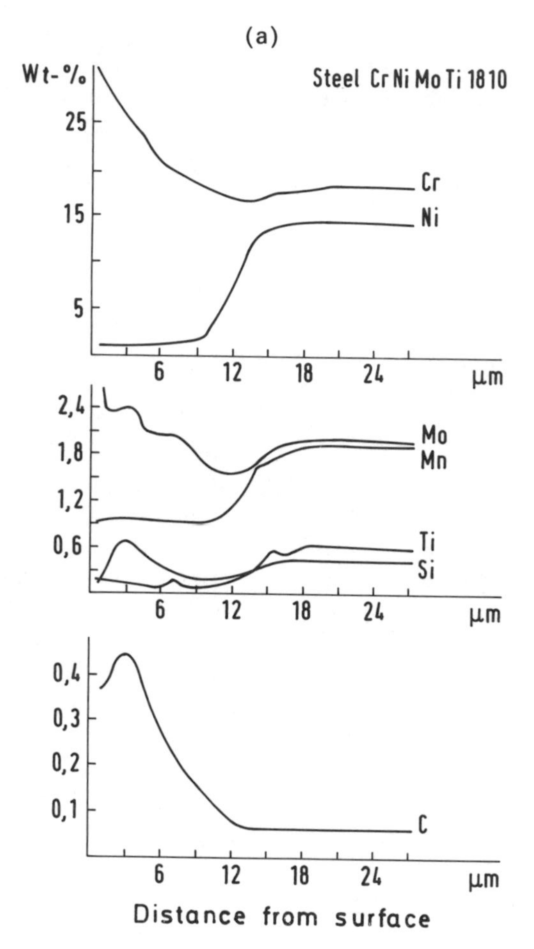

Figure 10.9. Effects of liquid lithium corrosion at 700°C on the steel X10 CrNiMoTi 18 10. (a) Concentration properties in the corroded layer.

The corrosion resistance of molybdenum and its alloys is also excellent in the heavier alkali metals, and the behavior is similar to that in sodium. Even the formation of a potassium molybdate, K_2MoO_4, as a corrosion product in the system K–O–Mo has been reported. Molybdenum is also one of the most compatible metals with molten lithium, as the very few experimental data indicate.

Tungsten seems to have the same good resistance against alkali metal corrosion as molybdenum. The group VI metals can, therefore, be applied to form protective layers on less compatible materials. Very thin protective layers of such metals are obtained by means of plasma spraying. It is also

(b)

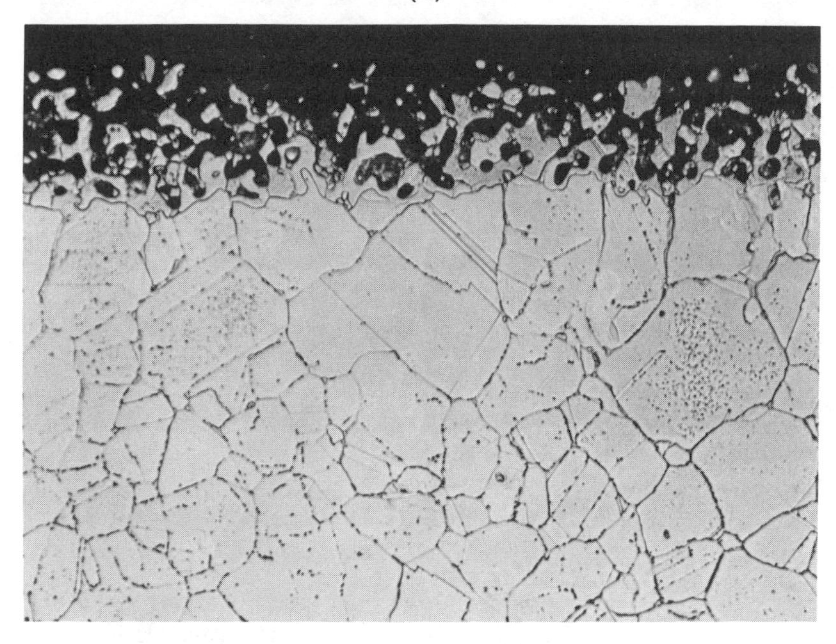

Figure 10.9 *(continued).* (b) Surface changes due to lithium corrosion.

reported that rhenium is fairly well compatible with alkali metals at high temperatures.

Braze alloys based on tin, copper, silver, and other precious metals are not suitable for liquid alkali metal service, since these metals are more or less soluble. A tin-based brazing of sheet material has been found heavily attacked by sodium–potassium eutectic alloy even at room temperature. The brazing failed after some minutes of exposure. High-temperature brazing alloys based on nickel powder as the main component are, however, sufficiently compatible with sodium and the heavier alkali metals. They are not applicable in lithium at high temperatures, because their main component is fairly soluble in this alkali metal. The minor components, boron and silicon, added as fluxing agents, may be leached out by alkali metals, especially by liquid lithium.

Ceramic materials resist attacks of sodium and the heavier alkali metals to a certain degree. The compatibility of the ceramic materials depends on their chemical stability as well as on their purity and degree of porosity.

Alumina (Al_2O_3) and thoria–yttria (ThO_2 with 7–15 wt% Y_2O_3) are

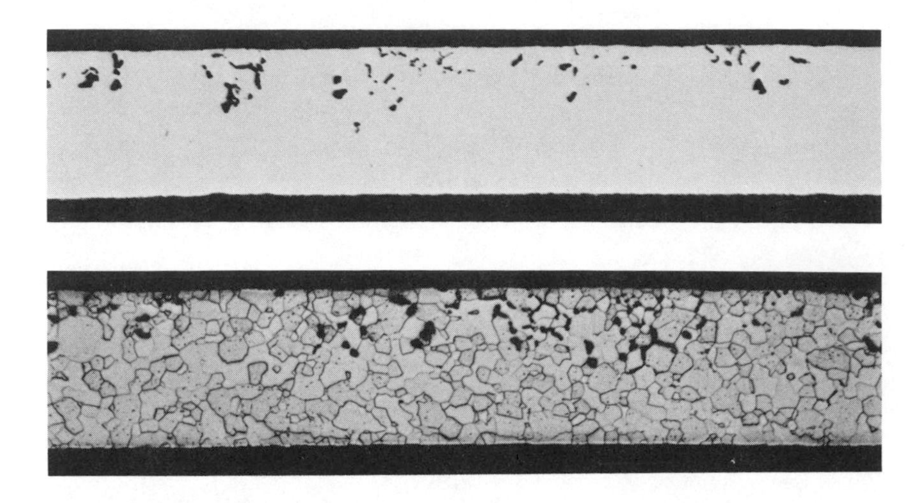

Figure 10.10. Cavity formation by leaching of oxygen out of vanadium due to lithium exposure at 700°C, cross sectional view of the specimen (top) after etching (bottom).

resistant to corrosion by sodium, potassium, and rubidium up to a temperature of 550°C, if the oxygen activity in the alkali metals is kept low. Lithium reacts with these ceramic materials as well as with several others. A complete reaction of alumina crucibles with liquid lithium has been observed at a temperature of 400°C. Thoria–yttria ceramic serving as solid electrolyte reacts with lithium with the formation of a black, thin surface

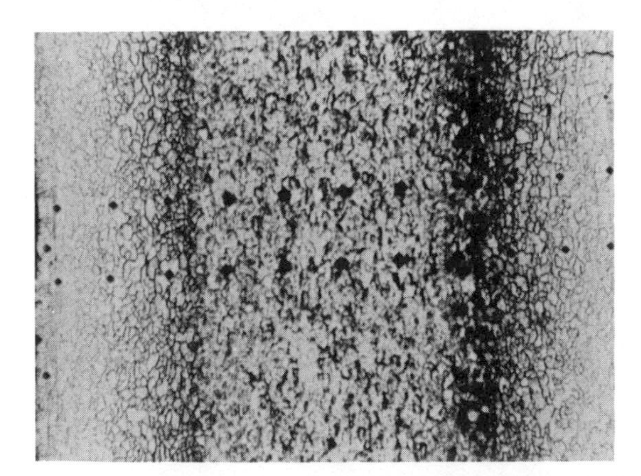

Figure 10.11. Uptake of nitrogen and formation of hardened layers containing precipitated nitrides of vanadium and niobium in a lithium-exposed specimen of V–3Ti–5Nb–10Cr (700°C) (with permission of VCH Verlagsgesellschaft, Weinheim, Germany).

layer. This layer seems to protect the oxygen meter probes made of this material against further attack by the molten metal.

Uranium dioxide serving as a nuclear fuel in fast breeder reactors may come into contact with sodium, if a fuel element cladding failure occurs. Chemical reactions of UO_2 with sodium containing dissolved oxygen have been extensively studied,[21] and the compound Na_3UO_4 has been identified as a reaction product. The formation of this compound in fuel elements gives rise to a swelling of the fuel. Small defects in the cladding tubes may grow due to such swelling, because the fuel puts high stresses on the tubes.

Uranium carbide fuel seems to be compatible with sodium up to temperatures of about 800°C. The stoichiometric uranium carbide and phases richer in uranium do not react with the alkali metal. Uranium carbides with higher carbon content, however, lose carbon when exposed to the liquid metal. A carbon transfer between uranium carbide and austenitic steel occurs, in which the stainless steel serves as a carbon sink. The source of carburization seems to be UC_2 mixed with UC, while mixtures of UC with U_2C_3 cause only weak carburization of the steels.

10.5. Corrosion and Mass Transfer

In all technical systems containing materials in contact with liquid alkali metals in which solution of substitutional and interstitial alloying elements occurs, the dissolved species are transported by means of the liquid phases. The first step of this transport of elements is the migration of the dissolved elements or compounds through the boundary layers into the free flow of the liquids. In the case of flowing liquid metals, further transport takes place by convection. In stagnant fluids, diffusion is the transport mode.

If the systems contain not only a source for the dissolution of elements but also a sink, the elements leached out from one material are transferred to another material, which contains this element at a lower level of chemical activity. The lower activity may be caused by a lower local temperature or by a different chemical composition, such that the chemical activity of this particular element is decreased. In the first case, thermal mass transfer occurs; in the second, chemical mass transfer may be observed.

The liquid alkali metals flow through sections of higher or lower temperatures in dynamic systems. Corrosion occurs, as described, in the high-temperature sections of circuits. Corrosion leads to dissolution of elements and to increased concentrations of dissolved species in the alkali metals.

Due to the temperature-dependent solubilities of the elements, their chemical activities may increase when the concentrations and the temperatures decrease. In an isothermal tube of high temperature, concentration and chemical activity of dissolved elements increase in the same sense, if there is no interference among the several solutes.

The rates of dissolution become very small in regions with decreasing temperatures; thus there is nearly no further change in concentration. The chemical activities, however, become very high due to the lower solubilities. Precipitation of the solutes is, therefore, observed when the chemical activities of the dissolved metals reach values in the region of the activities of the alloying elements in the solid materials. These activities are generally below unity. Precipitation is observed, therefore, even at chemical activities well below unity. The formation of compounds often leads to activities orders of magnitude lower. The formation of particulates of compounds purifies the molten metals to a very high degree and enhances their capacity to dissolve metallic elements.

Chemical mass transfer occurs even in isothermal systems. Different chemical activities of the alloying elements in two materials are the driving forces for this type of mass transfer. A typical example has been observed in a hot trap applied for the purification of liquid lithium. The zirconium getter used in this trap has the tendency to form intermetallics with nickel and chromium and acts as a sink for these elements. The activities of the two metals in solution are kept at low levels due to the gettering by zirconium foil. The stainless steel container therefore suffers enhanced corrosion due to the preferential leaching of these elements. This causes the formation of large corrosion layers in which ferrite and cavities occur. The growth of the corrosion layers indicates that the corrosion rates have been enhanced. The surface zones of the zirconium sheets collect the nickel up to a concentration of 10 wt%, and also some chromium.

In a nonisothermal lithium circuit, magnetic corrosion products are transferred from the corroded material in the hot leg to the cold part of the loop. If there is no provision made to trap the magnetic corrosion products, the particulates tend to precipitate in the magnetic fields of the pump and the flow meter, thus blocking the whole system within a short period of time ($\sim$500 h).

The precipitation behavior of different elements has been studied in sodium loops. The sodium leaches out chromium from austenitic steels in the high-temperature positions of the loops. The dissolved chromium is transferred by the flowing sodium to positions of lower temperatures. Chromium deposits in zones not far from the positions of dissolution, and hence the transfer is restricted to short distances. Precipitated chromium is never detected in the elemental form in the downstream regions of sodium loops. The element can, however, be found to cover wide areas of material surfaces in the cooler regions, deposited as sodium chromite, $NaCrO_2$, in typical layers of hexagonal crystals. In still cooler parts of the tubes of sodium circuits, even chromium carbides can be detected as a coating on the walls.

Nickel and manganese precipitate in the form of metallic particles. They tend to accompany each other when they are precipitated. These particles often have the form of cubic crystals, and are spread over the layers of sodium chromite or carbide. The ratio of manganese to nickel content in the crystals is higher than in the austenitic steels. The metallic deposits are also on the surfaces of tubes in the cold parts of a loop, even of the cold traps. The manganese content of the particulates is higher in the low-temperature region than at higher temperatures.

10.6. Transfer of Activated Corrosion Products

Mass transfer in the primary circuits of a fast breeder reactor leads to the buildup of radioactivity outside the core region. This is due to the activation of the cladding materials through neutron reactions:

$$^{50}\mathrm{Cr}(\mathrm{n}, \gamma) \rightarrow {}^{51}\mathrm{Cr}, \qquad \tau_{1/2} = 27.8\ \mathrm{d} \tag{3}$$

$$^{59}\mathrm{Co}(\mathrm{n}, \gamma) \rightarrow {}^{60}\mathrm{Co}, \qquad \tau_{1/2} = 5.3\ \mathrm{a} \tag{4}$$

$$^{54}\mathrm{Fe}(\mathrm{n}, \mathrm{p}) \rightarrow {}^{54}\mathrm{Mn}, \qquad \tau_{1/2} = 314\ \mathrm{d} \tag{5}$$

$$^{58}\mathrm{Ni}(\mathrm{n}, \mathrm{p}) \rightarrow {}^{58}\mathrm{Co}, \qquad \tau_{1/2} = 71\ \mathrm{d} \tag{6}$$

All the radioactive isotopes formed by reactions (3)–(6) have a relatively long lifetime, and they emit strong radiation. Corrosion of the cladding materials causes the dissolution of steel elements, radioactive elements among them. Liquid sodium transfers them along the temperature gradient.

The behavior of activated isotopes, their dissolution in liquid sodium, and their transfer in a sodium loop have been extensively studied by Brehm and Anantatmula.[22] The results of this work at Hanford indicate that ^{54}Mn is preferentially released from steel AISI 316 at all test temperatures and uninfluenced by variations in the oxygen content. Fast transport of manganese from the steel surface into the bulk sodium flow and high diffusion rates in the solid phase seem to be the reasons for the superstoichiometric release. The high diffusion rate might be favored by the vacancies originating from the enhanced chromium leaching. The growth of the surface zone depleted in manganese has been observed to be faster than is the case for other elements. Figure 10.12 shows the measured concentration profiles of manganese in a sodium-corroded surface of steel AISI 316.

The isotope ^{60}Co, on the other hand, is preferentially retained in the same steel. The surfaces of specimens of steel AISI 316 contain 20 to 30% more cobalt than the matrix. This element behaves like iron, which is also substoichiometrically released.

Thus, manganese is transferred from the fuel element clads through the hot piping into the heat exchanger, where the temperature decreases. This cooling causes the enhanced precipitation of manganese. Manganese tends to crystallize together with nickel, which is the main component of the deposits. The deposition of the radioactive ^{54}Mn on the tube walls of the heat exchanger causes maintenance problems if the heat exchanger fails. Cobalt, on the other hand, tends to be fixed in the hot piping of the primary circuit. Its radioactivity is not transferred outside the shielded area of the primary circuit.

^{51}Cr may contribute to the activation of the heat exchanger tubes. It is preferentially leached out of high temperature components, such as manganese. However, its precipitation behavior differs completely. Chromium deposits in the nonmetallic state. Compounds like carbides or oxides form layers on the tube walls. The compound $NaCrO_2$ can be detec-

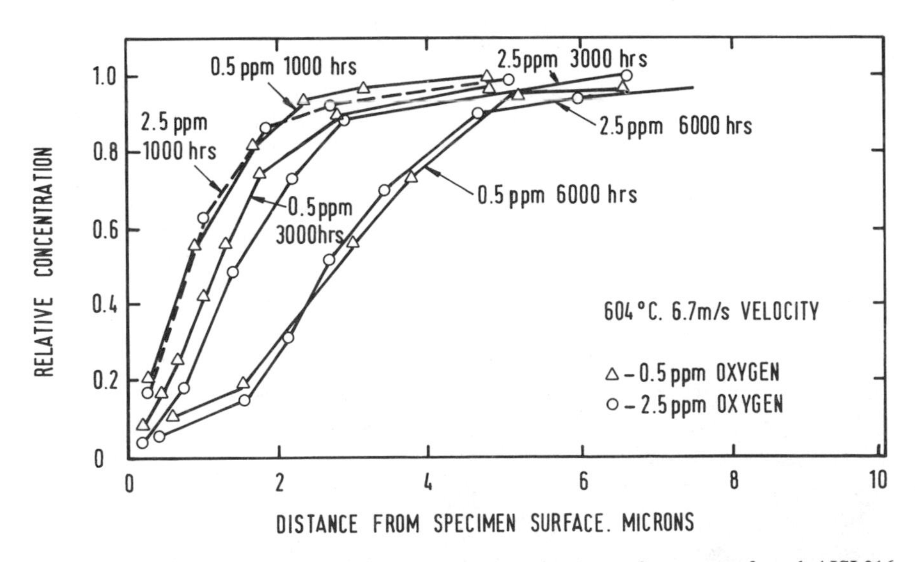

Figure 10.12. Concentration profiles of manganese in the surface zone of steel AISI 316 exposed to sodium at 604°C and a velocity of 6.7 m/s (after ref. 22; reproduced with the permission of Plenum Press, New York).

ted, if the removal of sodium is performed without the use of hydrolyzing processes.

The activation of the primary circuit is a serious problem. It may be minimized by reducing corrosion. Reduction of the oxygen concentration does not suppress the manganese release. Thus, reduction of the maximum temperature or trapping of activated elements might be necessary to solve the problem. Trapping is performed by means of absorbing metal foils, which retain manganese in the hot part of the sodium system.

The behavior of the element zinc in a fast reactor primary circuit has been studied in the Prototype Fast Reactor at Dounray.[23] The analyses of sodium samples from several reactors have indicated that the isotope ^{65}Zn is present in those sodium systems. The source of this radioactive isotope should be a content of some parts per million of the element in the primary coolant, into which it might come due to extraction from structural materials. A theoretical model developed on the basis of a knowledge of the sodium–zinc binary system gives an explanation for the complicated distribution of the element in such a sodium circuit. Zinc tends to be concentrated at the liquid–solid and liquid–gas boundaries of the sodium due to its limited solubility.

The concentration of zinc at the liquid–solid boundary causes its adsorption on the steel surfaces and its diffusion into the steel matrix as well. Evaporation of sodium surface layers, in which zinc might be accumulated, occurs at the sodium–gas interface. Sodium vapor with high zinc content condensing on component wall surfaces transfers zinc through the cover gas. The adsorption of zinc on the material surfaces occurs in this case, too. It migrates into the matrix of the materials due to diffusion processes if the temperatures are sufficiently high. The "plate-out" of zinc in sodium circuits may cause the embrittling of several iron-based alloys, since this metal belongs to the group of classical embrittlers.

Zinc cannot be determined in sodium by means of the distillation method, as it gets lost in the distillation process.

10.7. Corrosion Modeling

A model describing the dissolution as well as the precipitation of material in a liquid alkali metal circuit has not yet been developed. Some basic considerations, however, have been published. Several attempts have been made concerning the best-known system, sodium–austenitic stainless steel. From experimental results and from the examination of precipitated materials, Kolster[24] concluded that there is a threshold oxygen activity in sodium beyond which corrosion is dependent on oxygen concentration.

Sodium chromite, $NaCrO_2$, is stable enough to play its role in the corrosion reactions if the chemical activities of both oxygen and chromium are high enough. The precipitation results in an orientated growth of the crystals with their hexagonal axes perpendicular to the substrate surfaces. This can only occur if atoms or ions are the components of the crystals growing from nuclei present on the surfaces. The boundary of the crystals by $|10\bar{1}4|$ planes can be explained by the fact that Na^+ and CrO_2^- ions take part in the formation of the deposits.

$NaCrO_2$ cannot be observed at low oxygen concentrations in sodium. Chromite ions do not form in the heated zone of a system filled with pure sodium. In the cooler region, where the stability of chromite is increased, nuclei for crystallization are not available. In that case, chromium is partly deposited as a carbide, which cannot be a primary corrosion product. Carbon present in sodium acts as a getter for chromium, thus reducing the chemical activity of this metal and enhancing its dissolution rate.

Though corrosion of iron by alkali metals is also dependent on the oxygen activity in the melt, no solid ferrite has ever been detected in sodium loops. Kolster assumes that clusters of oxygen and sodium ions with a high electron density at their boundaries are present even at low concentrations of both ions. It might be that iron lowers its free energy by forming a complex with such a cluster, in which it is fixed at the boundary. The complex formation is a process which favors the higher solubility of the solid metal in the molten metal. This assumption is consistent with the observed solubility of iron in sodium containing different levels of oxygen concentrations.(3) Only a very small fraction of the dissolved oxygen has to be present in the active state, in which it is part of the cluster. It seems that the clusters become unstable at lower temperatures and are converted into molecular sodium oxide. Under these conditions, iron will be precipitated in the metallic state.

It is obvious from several experimental results on the steady-state corrosion of metals by liquid sodium that iron dissolution has to be the rate-controlling process in the corrosion of austenitic steels at high temperatures. Therefore, Thorley(25) has done studies to evolve a model of iron corrosion in sodium with an oxygen content far below the saturation level. He found that the iron corroded by liquid sodium with ~ 80 wppm oxygen did not consume any of the oxygen. He could not detect any oxidized iron species in the hot and cold tubes of the mild steel loop, and there were no particulates in filters or in separators. The corroded surfaces appeared metallic. They were not covered by any insoluble material or corrosion products. Evidently, a dissolution process is involved. Oxidation has, therefore, to be excluded as a possible mechanism, though oxygen is involved in the corrosion of iron in sodium. It seems that the assumption of an influence of the oxygen concentration on the chemical activity of the dissolved iron can help to explain the dependence of the corrosion rates on the oxygen activity. Also, Thorley has concluded that oxygen may only act as a carrier of iron, referring to Kolster's model.

Models for the corrosion of alloys in other alkali metals have not yet been developed to a state which allows their discussion.

10.8. Recommendations on Materials for Alkali Metal Systems

Several metallic and nonmetallic materials are more or less compatible with liquid alkali metals. They may be used as container and structural materials in liquid alkali metal technology even in the high-temperature range. Table 10.1 gives a survey of the compatibility of several metallic and nonmetallic materials with the alkali metals. Recommendations concerning the applicability of materials for service in alkali metal technology can be drawn from the information collected in this table.

The upper temperature limits are based on corrosion by flowing alkali metals at relatively high flow velocities. They are not valid for systems with stagnant alkali melts in which transport by means of convection and deposition may be neglected. Refluxing alkali metals do not possess the capacity to transfer dissolved materials. Thus the corrosion in heat pipes and similar systems is less serious than in loops and allows the use of several materials for the structure of such devices even for use at still higher temperatures. Components which are constructed with thick-walled tubes or vessels can also be operated at higher temperature than the limit given in the table, if the mass transfer does not cause additional problems.

The high-temperature technology of alkali metals which may be entered in the near future to make use of the heat transfer capacity of these liquid metals in the temperature range from 700 to 1200°C has not yet been studied thoroughly. The construction of such heat transfer circuits has to take into account the fact that material loss due to corrosion may limit the lifetime of the tubes exposed to alkali metals at such high temperatures. It seems that some superalloys based on nickel, which contain molybdenum and cobalt, are less sensitive to alkali metal corrosion than would be expected from their nickel content. These alloys, which have also very good high-temperature strength, may be used for this range of alkali metal applications. Their corrosion rates under the conditions of such applications are still to be evaluated.

The corrosion of materials in the NaK eutectic alloy can be considered to be very similar to sodium corrosion. The corrosion of materials in con-

Table 10.1. Compatibility of Materials with Alkali Metals

Material	Compatible with alkali metal up to (°C)				Factors influencing compatibility
	Li	Na	K	Rb and Cs	
Mg alloys	n.c.[a]	n.c.	300	300	Metal solubility, oxygen exchange
Al alloys	n.c.	350	400	450	Metal solubility
Cu alloys	300	400	400	400	Metal solubility
Ag and its alloys	n.c.	n.c.	n.c.	n.c.	High metal solubility
Au and its alloys	n.c.	n.c.	n.c.	n.c.	High metal solubility
Zn coatings	n.c.	n.c.	n.c.	n.c.	High metal solubility
Pb and its alloys	n.c.	n.c.	n.c.	n.c.	Very high metal solubility
Sn and its alloys	n.c.	n.c.	n.c.	n.c.	Very high metal solubility
Fe	500	700	700	700	Nonmetallic impurities
Low-alloy steels	500	700	700	700	Nonmetallic impurities
Ferritic steels	500	700	700	700	Nonmetallic impurities
High-Cr steels	500	700	700	700	Nonmetallic impurities
Austenitic steels	450	750	750	750	Nonmetallic impurities
Ni alloys	400	600	600	600	Flow velocity
Mo alloys	1000	1000	1000	1000	Nonmetallic impurities
W alloys	1000	1000	1000	1000	Nonmetallic impurities
Ti alloys	700	700	700	700	Nonmetallic impurities
Zr alloys	700	700	700	700	Nonmetallic impurities
V alloys	700	700	700	700	Nonmetallic impurities
Nb alloys	700	700	700	700	Nonmetallic impurities
Ta alloys	700	700	700	700	Nonmetallic impurities
Sintered Al_2O_3	350	500	500	500	Thermomechanical action
stab. ZrO_2/CaO	350	350	350	350	Intergranular corrosion
stab. ThO_2/Y_2O_3	400	550	550	550	Intergranular corrosion
Glass	n.c.	250	250	250	Chemical reaction
UO_2		750			Excess of oxygen
UC		750			Nonmetallic impurities

[a] n.c. = not compatible.

tact with the lithium–lead eutectic alloy, however, is more severe than the lithium corrosion at the same temperature and flow velocity. Corrosion rates with the $Li_{17}Pb_{83}$ alloy are up to five times faster than with lithium itself. The stronger corrosion effects are due to the high solubility of several metals in liquid lead. The lithium content of the alloy enhances the wetting of the materials and supports in this way the effect of lead.(28)

10.9. Corrosion Testing Methods

Corrosion or compatibility tests of materials in molten alkali metals should be related to the parameters of the alkali metal application.(26,27) In simple cases in which the molten metals are used in stagnant and more or less isothermal systems, simple testing devices may be used. The materials to be tested are employed as construction materials for closed capsules which contain the molten metals. The capsules are heated to the temperature of interest and kept under the desired conditions for different periods of time. The choice of the exposure times should be so made as to give data useful for the extrapolation of the corrosion effects to the duration of the application. The capsules are filled and mounted inside an inert-gas glove box (see Chapter 4).

The double-chamber capsule permits the gettering of alkali metals with zirconium or titanium sponge or foils before the corrosion tests. Capsules may be equipped with provisions to introduce impurities such as gas or condensed matter. The capsule with oxygen meter generates data describing reactions like oxygen exchange. Isothermal mass transfer can be studied by means of bimetallic capsules. Instead of bimetallic capsules, crucibles containing sheets of the second metal are also in application.

A step towards the application of liquid metal circuits is the autoclave with a purification unit. The purification unit is a very small circuit with a cold trap. This circuit provides the test chamber with freshly purified alkali metal melt. The tests in this kind of device are not purely diffusion controlled, since convection contributes to corrosion and mass transfer.

Thermoconvection loops are circuits containing a heater and a cooler which generate temperature gradients. The liquid metal flows through the loop due to the density differences in the two legs. Flow velocities of some centimeters per second are generated by the thermoconvection. The force of this flow is not sufficient to let the liquid metals pass through cold traps. The alkali metal chemistry of these loops is, therefore, not well defined.

Alkali metal loops with forced circulation provide high flow velocities, high purity, and the opportunity to apply in-line analytical equipment. The corrosion parameters in such loops are well defined. Studies of corrosion and deposition can be made at the same time. This enables the experiment to provide data useful for modeling the corrosion effects. If large temperature gradients are necessary, loops with a figure-eight shape are constructed which contain a sodium–sodium heat exchanger to economize on energy consumption. Cold traps and sampling stations are placed in bypass lines of the cold leg of the loops. Test sections, hot traps, in-line monitors, and foil equilibration devices are connected with the hot leg. The test sections may also have provisions which allow corrosion tests under the influence of mechanical stresses.

Figure 10.13 shows an example of a creep-rupture test section, which is connected with the hot leg of a sodium loop. The provisions to measure the constant load and the growing strain of the specimens are below the test

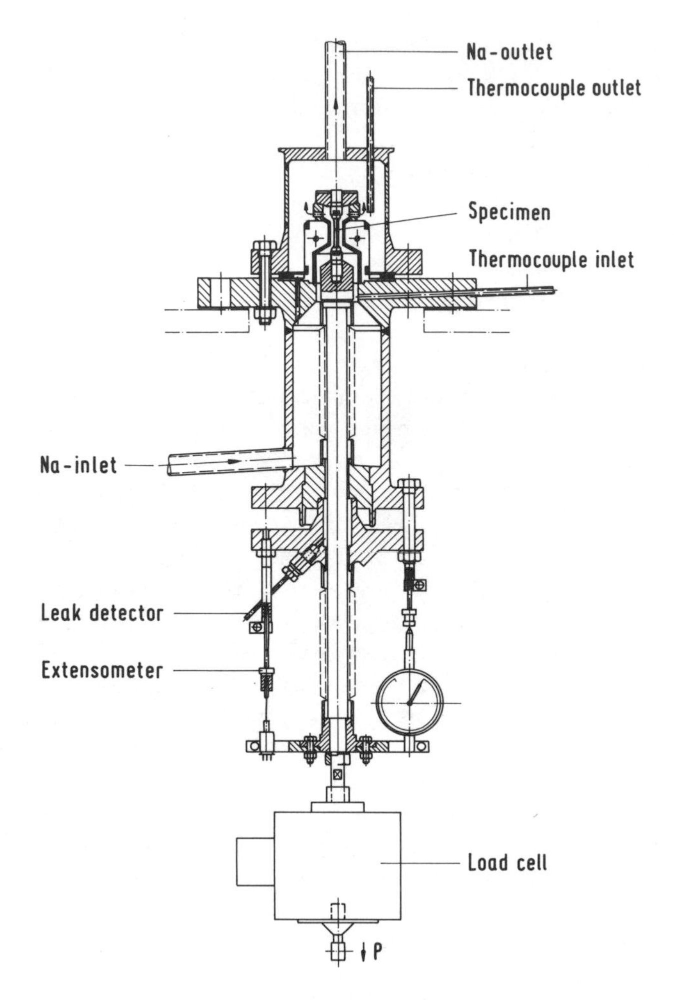

Figure 10.13. Test section for creep-rupture tests in flowing sodium, as used in the Karlsruhe Nuclear Center.

section. Bellows close the test section and allow the load to generate a creep strain. The flow velocity in the gap between the specimens and the housing reaches the order of 1 to 3 m/s. Such test sections in a modified design are in use for low-cycle fatigue tests in flowing sodium.

The thermal expansion of the liquid sodium in the temperature range between the filling and the testing temperature makes necessary the provision of an expansion tank, which has to be placed at the highest spot of the sodium system. It serves also for the loading of a gas pressure on the liquid metal surface. Gas analytical equipment placed in the cover gas volume in this expansion tank allows the measurement of impurities which might be evaporated.

Parallel arrangement of the corrosion test sections allows the study of the influence of the test temperature on corrosion kinetics. Test sections in series, however, are useful for the examination of the mass transfer along a loop with temperature gradient.

Even glass apparatus may be used for corrosion tests in the heavier alkali metals at temperatures below 300°C. Glass is compatible at this temperature with the alkali metals heavier than lithium. Its application allows the construction of simple and cheap loops for studies of materials which suffer considerable alkali metal corrosion effects at this low temperature level. These loops made of glass and equipped with a simple electromagnetic pump have been used for several studies at Nottingham University.

The application of classical methods as well as of modern chemical surface analytical instruments is necessary to evaluate the corrosion effects. Weight and thickness measurements of the exposed specimens are used to determine corrosion rates. Metallographic techniques give additional information on internal corrosion effects which are sometimes more pronounced than weight losses. Precipitation of compounds, phase changes in spots and in layers, and embrittlement or losses of strength are indicated by means of these methods. The formation of cavities or deposition layers is shown as well.

The microprobe is employed to measure changes of the chemical composition in corrosion layers. The resolution of the microprobe is, however, often insufficient to locate exactly the changes or to determine concentration gradients. Glow discharge optical spectrometry[20] is an excellent method for measuring the gradients of the concentrations of various alloying elements in layers of 0.005 to 0.100 mm thickness. Auger electron spectrometry sputters much slower than the glow discharge and is used, therefore, to estimate the concentration profiles close to the surfaces. These two methods have also the capacity to detect changes in concentrations of nonmetallic elements such as carbon or nitrogen. Scanning Auger electron spectrometers have the advantage of combining the Auger analyses with a high-resolution microprobe.

The combination of several evaluation techniques enables us to combine several corrosion phenomena to get an overall picture. Losses of wall thickness, formation of internal corrosion layers of changed chemical composition, structure, and mechanical strength, and any other effects have to be considered in order to get full information on corrosion damage. Thus, predictions on the compatibility of structural materials with alkali metals and on the influence of the corrodants on the lifetime of circuit components in alkali metal systems may be based on the knowledge of the corrosion phenomena and their dependence on time and parameters.

References

1. H. Leyerzapf, *Werkst. Korros. 36*, 88–96 (1985).
2. A. W. Thorley and C. Tyzack, in: *Liquid Alkali Metals*, British Nuclear Energy Society, London, 1973, pp. 257–273.
3. S. P. Awasthi and H. U. Borgstedt, *J. Nucl. Mater. 116*, 103–111 (1983).
4. H. U. Borgstedt, in: *2nd Internat. Conf. on Liquid Metal Technology in Energy Production* (J. M. Dahlke, Ed.), National Techn. Information Service, Springfield, Va., 1980, (CONF-800401 P1), Vol. 1, 7-1.
5. C. F. Cheng and W. E. Ruther, *Corrosion (NACE) 28*, 20–22 (1972).
6. H. U. Borgstedt, *Corros. Sci. 11*, 89–105 (1971).
7. J. E. Selle, in: *Internat. Conf. on Liquid Metal Technology in Energy Production* (M. H. Cooper, Ed.), National Techn. Information Service, Springfield, Va., 1976, (CONF-760503 P2), Vol. 2, pp. 453–461.
8. A. W. Thorley and C. Tyzack, in: *Alkali Metal Coolants*, International Atomic Energy Agency, Vienna, 1967, pp. 97–118.
9. P. Roy, G. P. Wozadlo, and F. A. Comprelli, in: Proc. of the Intern. Conf. on Sodium Technology and Large Fast Reactor Design, Report ANL-7520, 1969, Part 1, pp. 131–142.
10. B. J. Shaiu, P. C. S. Wu, and P. Chiotti, *J. Nucl. Mater. 67*, 13–23 (1977).
11. P. G. Gadd and H. U. Borgstedt, in: *Liquid Metal Engineering and Technology*, British Nuclear Energy Society, London, 1984, Vol. 1, pp. 107–111.
12. C. Bagnall and D. C. Jacobs, USERDA Report WARD-NA-3045-23, 1975.
13. M. G. Barker, S. A. Frankham, P. G. Gadd, and D. R. Moore, in: *Material Behavior and Physical Chemistry in Liquid Metal Systems* (H. U. Borgstedt, Ed.), Plenum Press, New York, 1982, pp. 113–120.
14. K. Natesan and T. F. Kassner, *J. Nucl. Mater. 37*, 223–235 (1970).
15. H. U. Borgstedt, G. Frees, and H. Schneider, *Nucl. Technol. 34*, 290–298 (1977).
16. W. T. Lee, *Nucl. Applic. Technol. 7*, 155–163 (1969).
17. R. S. Fidler and M. J. Collins, *Atomic Energy Rev. 13*, 3–50 (1975).
18. H. U. Borgstedt, G. Frees, and G. Drechsler, to be published.
19. C. Tyzack and A. W. Thorley, in: *Ferritic Steels for Fast Reactor Steam Generators*, Vol. 2, British Nuclear Energy Society, London, 1978, pp. 241–257.
20. H. Schneider and H. Schumann, Report KfK 2009, Kernforschungszentrum, Karlsruhe, 1974.
21. C. C. Addison, M. G. Barker, R. M. Lintonbon, and R. J. Pulham, *J. Chem. Soc. (A)*, 2457–2459 (1969).
22. W. F. Brehm and R. P. Anantatmula, in: *Material Behavior and Physical Chemistry in Liquid Metal Systems* (H. U. Borgstedt, Ed.), Plenum Press, New York, 1982, pp. 193–204.
23. E. H. Voice, in: *Liquid Metal Engineering and Technology*, British Nuclear Energy Society, London, 1984, Vol. 1, pp. 259–263.
24. B. H. Kolster, *J. Nucl. Mater. 55*, 155–168 (1975).

25. A. W. Thorley, in: *Material Behavior and Physical Chemistry in Liquid Metal Systems* (H. U. Borgstedt, Ed.), Plenum Press, New York, 1982, pp. 19–36.
26. H. U. Borgstedt and G. Frees, *Werkst. Korros. 30*, 91–100 (1979).
27. R. L. Klueh and J. H. DeVan, in: *Handbook on Corrosion Testing and Evaluation* (W. H. Ailor, Ed.), John Wiley and Sons, New York, 1971, p. 405.
28. D. L. Olson, G. N. Reser, and D. K. Matlock, *Corrosion (NACE) 36*, 140–144 (1980).

11

Chemical Reactions in Alkali Metals

Different nonmetallic or metallic elements dissolved in liquid alkali metals are often able to react with each other. The products of such reactions are often ternary compounds such as, for instance, the double oxides already described in Chapter 5. The chemical reactions of elements dissolved in liquid alkali metals sometimes produce compounds which are uncommon and only stable in the presence of excess alkali metal.

11.1. Reactions of Oxygen and Hydrogen in Alkali Metals

Reactions of alkali metal hydrides with oxides in liquid sodium systems are of technological importance and are, therefore, extensively studied. The equilibrium

$$NaH + Na_2O \rightleftarrows NaOH + 2Na \quad (1)$$

has been investigated using electrochemical meters for oxygen and hydrogen. Measurements of the hydrogen partial pressure over the reaction mixtures in excess sodium have been used to determine the equilibrium concentrations of hydride. The equilibration of the reactants at temperatures between 350 and 550°C in capsules of pure nickel is the experimental basis of a study of the Na–NaOH–Na_2O–NaH system at the Argonne National Laboratory.[1] In fair agreement with thermodynamic predictions,[2] the experimental data of this study indicate that sodium hydroxide decomposes at temperatures below 412°C to form NaH and

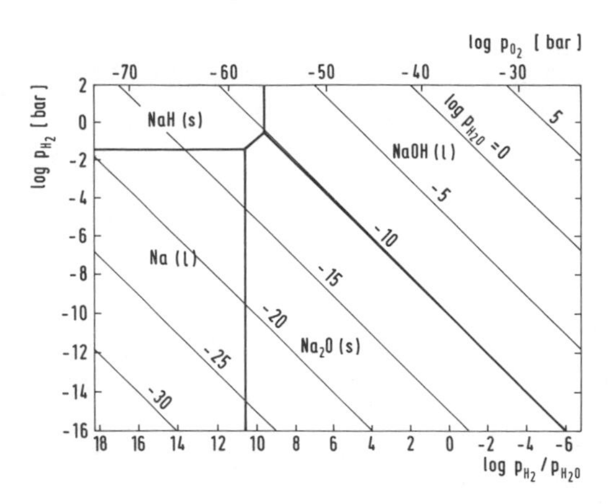

Figure 11.1. Thermochemical data for the condensed phases in the sodium–oxygen–hydrogen system at 327°C (after ref. 2; reproduced with the permission of Plenum Press, New York).

Na_2O. The hydroxide forms at high temperatures and high chemical activities of hydrogen and oxygen. Figures 11.1 and 11.2, taken from ref. 2, demonstrate the ranges in which the phases NaOH, NaH, and Na_2O exist in contact with liquid sodium at 327 and 527°C. The equilibria corresponding to equation (1) are also observed in the liquid eutectic alloy of sodium and potassium saturated with oxygen or even containing low amounts of oxygen.

Lithium hydroxide is not stable in contact with an excess of the liquid metal. Thus, oxygen and hydrogen do not react when dissolved in liquid

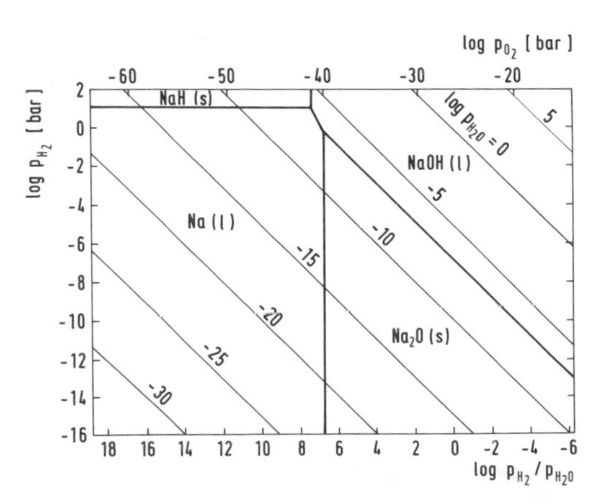

Figure 11.2. Thermochemical data for the condensed phases in the sodium–oxygen–hydrogen system at 527°C (after ref. 2; reproduced with the permission of Plenum Press, New York).

lithium. The importance of this fact for fusion reactor technology is obvious. Tritium formed by nuclear reaction cannot be captured by lithium oxide present in the liquid metal.

The thermochemical and kinetic aspects of the formation of sodium hydroxide in dilute solutions in the liquid metal are critically reviewed in a recent review paper.[3] The following equation for the solubility of sodium hydroxide in sodium is taken from this article:

$$\log[x_{NaOH}]_s(\text{wppm}) = 2.73 - \frac{3000}{T(\text{K})} \tag{2}$$

The stability of alkali hydroxides in contact with the liquid metals seems to increase with the atomic weight of the alkali metal. Thermochemical estimations based on the work of S. A. Jansson indicate the relatively high stability of potassium hydroxide in contact with the liquid metal. The hydroxide is stable in liquid potassium from the melting point of the metal up to its boiling point at sufficiently high chemical activities of oxygen and of hydrogen.[4] The range of stability of the potassium hydroxide phase increases with increasing temperature, in agreement with the findings in the sodium–oxygen–hydrogen system. If the tendency shown by the series LiOH–NaOH–KOH continues on going to the higher atomic weight alkali metals, rubidium and cesium hydroxides should also be stable in contact with the metals.

The kinetic and thermodynamic aspects of the interactions of oxygen and hydrogen with an excess of sodium have been extensively studied by Smith and Whittingham.[5] They were able to show that the decomposition of sodium hydroxide proceeds according to the equations

$$2Na_{(l)} + NaOH_{(l)} \rightleftarrows Na_2O_{(s)} + NaH_{(s)} \tag{3}$$

and

$$Na_2O_{(s)} \rightarrow |O|_{\text{diss.}} \tag{4}$$

$$NaH_{(s)} \rightarrow |H|_{\text{diss.}} \rightleftarrows \tfrac{1}{2}H_{2(g)} \tag{5}$$

The first reaction was found to be fast, while the two others proceed slowly. The reaction

$$Na_{(l)} + NaOH_{(l)} \rightarrow Na_2O_{(s)} + \tfrac{1}{2}H_{2(g)} \tag{6}$$

contributes to the decomposition and is also a rapid process. The evolution of hydrogen as measured by means of partial pressure monitoring indicates that only a small part of the hydroxide is decomposed according to equation (6). The fraction of hydrogen released from mixtures of sodium metal with sodium hydroxide increases with increasing reaction temperature. Even at 500°C the decomposition is predominantly ruled by equations (3–5).

The reaction rates for the combination of hydrogen and oxygen in sodium have been measured using electrochemical meters for oxygen and hydrogen. The free energy of formation of the hydroxide in sodium has been calculated from equilibrium constants. The values are in close agreement with the equation given by Ullmann *et al.*[6]:

$$\Delta G^0_{OH}(\mathrm{J/mol}) = -(25{,}967 \pm 2681) - (28.762 \pm 4.136) \cdot T \qquad (7)$$

The results of equilibration measurements for the formation and decomposition of hydroxide in sodium are in fair agreement.

11.2. Reactions of Carbon and Oxygen in Alkali Metals

The technological importance of the sodium–carbon–oxygen system is due to the influence of oxygen concentrations on the solubility of carbon in the liquid metal. Several studies concerning carbon solubility in sodium conclude that there should be a tendency for the saturation concentrations of carbon to increase in the presence of dissolved oxygen. Thermodynamic analyses of the stability ranges of compounds in contact with metallic sodium have been performed by Migge.[7] Only the binary compounds sodium oxide and the metastable sodium acetylide (or the stable graphite) exist at low temperature and low potentials of both carbon and oxygen. Figure 11.3 (taken from ref. 7) exhibits the stability ranges of the compounds. It also shows that sodium carbonate can only coexist with the molten metal at high temperatures and concentrations of carbon and oxygen. Nevertheless, a saturation curve of the metastable carbonate has been calculated on the basis of some analytical values given in ref. (8). The saturation concentrations are expressed by the equation

$$\log[x_{\mathrm{Na_2CO_3}}]_s = 2.2459 - \frac{363.9}{T(\mathrm{K})} \qquad (8)$$

The flat gradient of the temperature function indicates that decomposition might be superimposed on the dissolution of the carbonate. The solubility of the unstable carbonate is quite high at low temperatures, and thus the carbonate concentration is relatively high at the temperature of the cold trap of a sodium system ($\sim$150°C). This is in agreement with findings con-

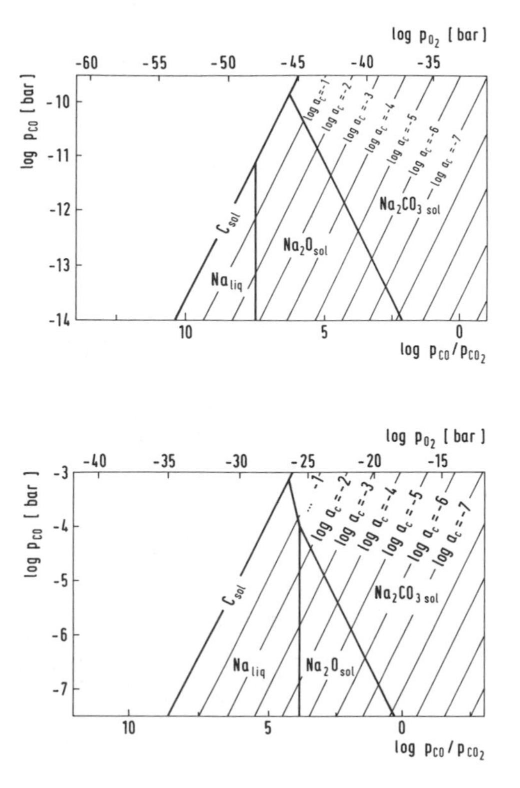

Figure 11.3. Stability ranges of carbonate in liquid sodium at 427°C (upper diagram) and 827°C (lower diagram) (after ref. 7; reproduced with the permission of Plenum Press, New York).

cerning the influence of cold-trap temperatures on the chemical activity of carbon in sodium loops. The analytically determined amounts of carbon are much higher than carbon in equilibrium with the precipitated element in the trap. The solubility of elemental carbon would result in much lower concentrations in the liquid metal. The measurement of the chemical activities of carbon at higher temperatures in such a nonisothermal sodium system, however, indicates that carbonate is unstable under the conditions existing in the liquid sodium in the hot part of such a circuit. The solutions behave as liquid metals containing elemental carbon or acetylide as the dissolved species.

Sodium carbonate has also been detected among the precipitated compounds in the cold traps of sodium loops, which contain in addition, other carbon-bearing compounds. Lithium carbonate is unstable in contact with metallic lithium over the whole range of temperatures in which lithium is liquid and even at very high concentrations of oxygen and carbon.[9] The stability of carbonates of the heavier alkali metals, particularly potassium, in contact with the molten metals might be higher, but data have not yet been published.

11.3. Reactions of Carbon and Nitrogen in Alkali Metals

Solutions of carbon in alkali metals react with nitrogen to form compounds containing both nonmetallic elements. Recently, the reaction of lithium acetylide with lithium nitride in excess liquid lithium has been studied. The formation of the cyanamide Li_2NCN has been observed. This stable compound has been isolated by removing the excess lithium metal by vacuum distillation.[10] The thermal stability of dilithium cyanamide is excellent, since it is the product of this reaction of acetylide with nitride even at a temperature of up to 700°C. Equation (9) indicates that this reaction liberates elemental lithium.

$$Li_2C_2 + 4Li_3N \rightleftarrows 2Li_2NCN + 10Li \tag{9}$$

There is some evidence that even lithium cyanide would be transformed into cyanamide if an excess of nitride is available in the liquid metal. The dissolution of cyanamide in liquid lithium influences the physical properties of the latter, such as the electrical conductivity, as well as the chemical activity of carbon and the partial pressure of nitrogen.

A thermochemical analysis of the lithium–carbon–nitrogen system has demonstrated that dilithium cyanamide can coexist with lithium metal over a wide range of temperatures and potentials of carbon and nitrogen in the liquid metal.[9] The standard free enthalpy of formation of Li_2NCN has been calculated to be $\Delta_f G^0 = -360$ kJ/mol.

The reaction of carbon with nitrogen is completely different in liquid sodium. Sodium cyanamide does not seem to be stable under the conditions existing in liquid sodium, where the solubility of nitrogen is extremely low. Such a compound is only formed in the presence of dissolved alkaline earth metals such as calcium, strontium, or barium. Thus, the reaction product is CaNCN, SrNCN, or BaNCN. The presence of alkaline earth metals, particularly barium, enables liquid sodium to dissolve more nitrogen than the pure metal. The higher nitrogen concentrations favor the formation of cyanamide. The nitrogen present even in saturated liquid sodium does not seem to be sufficient to allow the formation of sodium cyanamide.[11]

The product of the reaction of carbon with nitrogen in pure sodium at temperatures above 600°C is cyanide, NaCN. Cyanide formation causes an increase of the carbon solubility in sodium if the metal is in contact with an atmosphere of relatively high partial pressure of nitrogen. Sodium cyanide is also detected among the deposits in cold traps of sodium loops as one of the major carbon-bearing compounds.

The stability ranges of sodium cyanide in contact with the excess liquid metal at 527 and 727°C, taken from ref. 7, are shown in the diagrams of Fig. 11.4. Sodium cyanide should be stable in contact with the liquid metal according to these data. Its formation needs, however, high nitrogen potentials, which can only be obtained in the presence of alkaline earth metals. Reactions of carbon and nitrogen in the heavier alkali metals are not yet reported. It can be concluded that such reactions should occur only in the presence of dissolved alkaline earth metals.

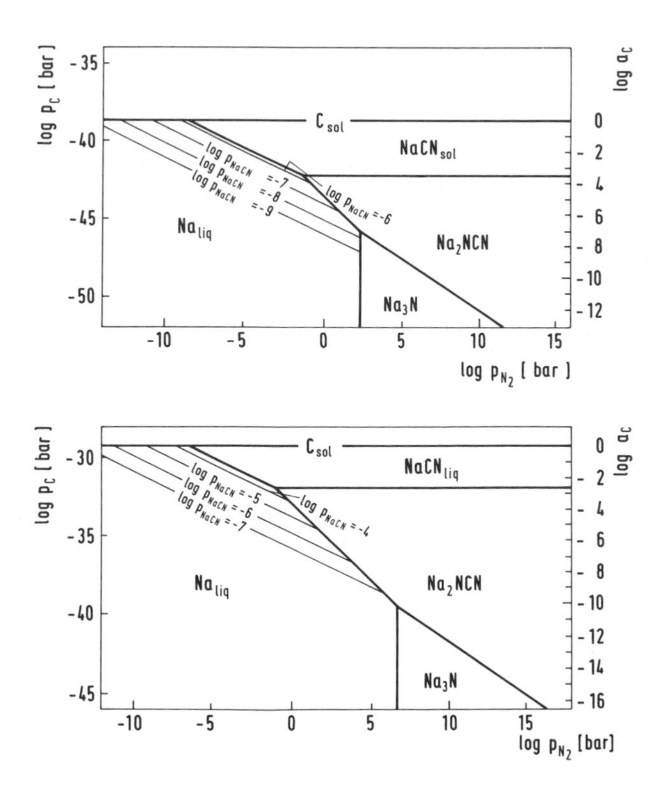

Figure 11.4. Stability ranges of cyanide in liquid sodium at 527°C (upper diagram) and 727°C (lower diagram) (after ref. 7; reproduced with the permission of Plenum Press, New York).

11.4. Reactions of Carbon or Nitrogen with Hydrogen in Alkali Metals

Carbon-bearing compounds, such as carbonates or cyanides, dissolved in liquid alkali metals, particularly in sodium, react with hydrogen or dissolved hydrides to form methane.[12] The stability of these compounds ensures that the hydrogenation does not occur at temperatures below 525°C, while sodium acetylide or elemental carbon dissolved in one of these liquid metals does not need such high reaction temperatures. Sodium or the eutectic sodium–potassium alloy promotes the hydrogenation of natural coal suspended in the molten metals through reaction with gaseous hydrogen introduced into the slurries.[13] These reactions also take place at relatively low temperatures of ~350°C. The molten alkali metals act as catalysts of the hydrogenation reaction, and they retain sulfur and other impurities. Therefore, the product of the hydrogenation of coal in alkali metals is a very pure methane gas.

11.5. Reactions of Nitrogen with Metals Dissolved in Alkali Metals

Addison *et al.*[15] have shown that the high solubility of nitrogen in sodium containing dissolved barium is caused by its reaction with this dissolved metal to form binary barium nitride, which is somewhat soluble in liquid sodium. Barium metal can be dissolved to concentrations of more than 5 at% at a temperature of ~300°C. Such solutions absorb nitrogen gas. Barium nitride formed by reaction with this absorbed nitrogen is soluble up to an atomic ratio of N:Ba = 1:4. The absorption of higher amounts of nitrogen causes the precipitation of barium nitride. The precipitated compound has the composition Ba_2N. Dissolved nitrogen seems to be present as nitride ion N^{3-}. The nitride ion is kept in solution by means of solvation, a process in which the composition Ba_4N seems to represent the limit of a stable atomic ratio in the liquid phase. Lower amounts of barium are not able to keep the N^{3-} ion in solution in liquid sodium. Thus, the precipitation of Ba_2N occurs whenever the concentration of nitrogen exceeds the above-mentioned limit. It is assumed that the solvation cluster $Ba_4\,|N|^{3-}$ has a tetrahedral structure.

The solid compound Ba_2N must contain free electrons. Its structure is of the anti-$CdCl_2$ type. Its formation in liquid barium–sodium alloy offers an alternative way to prepare this compound, normally prepared by means of the decomposition of the compound Ba_3N_2.

$$2Ba_3N_{2(s)} \xrightarrow{500°C} \tfrac{1}{2}N_2 + 3Ba_2N_{(s)} \qquad (10)$$

The resistivity method was used to study the dissolution of nitrogen in a sodium–barium liquid alloy. The precipitation of Ba_2N, which is completed on reaching its stoichiometric composition, leads to a decrease of the resistivity. The value of the resistivity reaches that of pure sodium when the precipitation of the compound is completed.[15] The alkaline earth metals calcium and strontium react in an analogous manner to barium, but these systems have not been studied to the same extent.

A case of a heterogeneous reaction of nitrogen with a metal in a liquid alkali metal is the formation of chromium nitride in sodium under a nitrogen partial pressure of the order of one bar at a temperature of ~700°C. The reaction product analyzed after cooling to room temperature has the composition Cr_2N. X-Ray diffraction, however, indicates that the reaction product is a mixture of chromium metal and the compound CrN. It seems that the first-mentioned compound is formed during the nitriding reaction at 700°C and decomposes during the cooling of the liquid metal.[16] Dissolved species might be involved in the reactions which cause the formation of solid compounds on the surfaces of specimens. As has been shown in Chapter 10, the corrosion reactions between stainless steel and liquid sodium result in the growth of corrosion layers in which chromium has been depleted. Thus, chromium needed for the reaction with nitrogen to form Cr_2N resp. CrN must be taken from the liquid phase. The chemical activity of chromium in sodium has to be high enough to favor the nitride formation. It is obvious that the formation of Cr_2N occurs at a considerably lower concentration level than that of the reactions of group II metals with nitrogen in liquid sodium.

11.6. Reactions in the Sodium–Chromium–Oxygen–Carbon System

In sodium systems carbon is an unavoidable impurity. Carbon potential in sodium increases with decrease in temperature. This is the reason for the precipitation of chromium-rich carbides in the cooler regions of a circulating sodium loop, which is followed by precipitation of sodium chromite. Under these circumstances, it is appropriate to consider the equilibrium

$$CrC_x + Na + 2\,|O|_{diss} \rightleftarrows NaCrO_2 + xC \qquad (11)$$

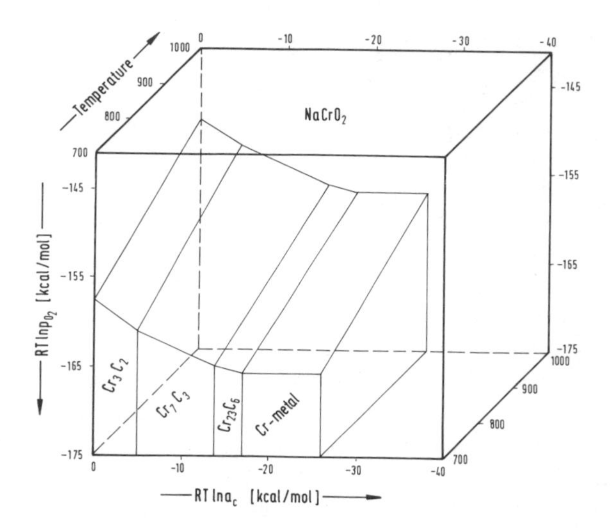

Figure 11.5. Phase stability diagram for the Na–Cr–O–C system (after ref. 17; reproduced with the permission of the authors).

The x in CrC_x depends on the carbon potential in the liquid metal. The phase stability diagram for this Na–Cr–O–C system has been recently computed.[17] The equilibrium phases are sodium, chromium, and sodium chromite at low carbon potentials. As the carbon potential increases, progressively $Cr_{23}C_6$, Cr_7C_3, and Cr_3C_2 precipitate. The equilibrium oxygen potential in the Na–CrC_x–$NaCrO_2$ systems has an analogous influence on the solubility of $NaCrO_2$ as the carbon potential for CrC_x. This implies that an increase in the amount of dissolved oxygen is necessary for the appearance of the $NaCrO_2$ phase. Figure 11.5 shows the stability ranges of the compounds in this quaternary system.

References

1. K. M. Myles and F. A. Cafasso, *J. Nucl. Mater. 67*, 249–253 (1977).
2. S. A. Jansson, in: *Corrosion by Liquid Metals* (J. E. Draley and J. R. Weeks, Eds.), Plenum Press, New York, 1970, pp. 523–560.
3. H. Ullmann, in: *Material Behavior and Physical Chemistry in Liquid Metal Systems* (H. U. Borgstedt, Ed.), Plenum Press, New York, 1982, pp. 375–386.
4. V. Ganesan and H. U. Borgstedt, to be published.
5. C. A. Smith and A. C. Whittingham, in: *Material Behavior and Physical Chemistry in Liquid Metal Systems* (H. U. Borgstedt, Ed.), Plenum Press, New York, 1982, pp. 365–374.
6. H. Ullmann, K. Teske, F. A. Kozlov, and E. K. Kiznekov, *Kernenergie 20*, 80–82 (1977).

7. H. Migge, in: *Material Behavior and Physical Chemistry in Liquid Metal Systems* (H. U. Borgstedt, Ed.), Plenum Press, New York, 1982, pp. 351–364.
8. H. T. Carmichael and S. A. Meacham, The Solubility of Sodium Carbonate in Sodium, USAEC Report APDA-184, 1968.
9. H. Migge, in: *2nd Intern. Conf. on Liquid Metal Technology in Energy Production* (J. M. Dahlke, Ed.), National Techn. Information Service, Springfield, Va., 1980 (CONF-800401-P2), Vol. 2, 18–19.
10. M. G. Down, M. J. Haley, P. Hubberstey, R. J. Pulham, and A. E. Thunder, *J. Chem. Soc., Dalton Trans.*, 1407–1411 (1978).
11. C. C. Addison, B. M. Davies, R. J. Pulham, and D. P. Wallace, in: *The Alkali Metals*, Spec. Publ. No. 22, The Chemical Society, London, 1967, pp. 290–308.
12. J. Jung, U. Buckmann, and R. Pütz, in: *Material Behavior and Physical Chemistry in Liquid Metal Systems* (H. U. Borgstedt, Ed.), Plenum Press, New York, 1982, pp. 265–273.
13. H. U. Borgstedt and J. Konys, *Chem. Ind. 36*, 404–406 (1984).
14. C. C. Addison, R. J. Pulham, and E. A. Trevillion, *J. Chem. Soc., Dalton Trans.*, 2082–2085 (1975).
15. C. C. Addison, G. K. Creffield, P. Hubberstey, and R. J. Pulham, *J. Chem. Soc., Dalton Trans.*, 1105–1108 (1976).
16. A. Marin, M. de la Torre, and H. U. Borgstedt, in: *2nd Intern. Conf. on Liquid Metal Technology in Energy Production* (J. M. Dahlke, Ed.), National Techn. Information Service, Springfield, Va., 1980, (CONF-800401-P2), Vol. 2, 13–20.
17. T. Gnanasekaran, G. Periaswami, V. Ganesan, S. Rajan Babu, and C. K. Mathews, *Intern. Conf. on Corrosion Science and Technology*, Calcutta, Febr. 21–23, 1985.

12

Reactions of Alkali Metals with Water

The violent reactions between alkali metals and water or steam have considerable technical importance. The alkali metals serve as heat transfer media in cooling circuits which transfer thermal energy to heat water or to produce steam. Such systems are used in fast breeder reactors, nuclear fusion reactors, or solar energy conversion plants. There are only thin metallic tube walls separating the two reactive media, in order to get an effective heat flux from the alkali metal to the pressurized water or steam. The water side of such steam generators is at much higher levels of pressure. Thus, there is the risk that water or steam may enter into the alkali metal-carrying pipes through leaks in the tube wall which might result if there is stress corrosion. Particularly austenitic stainless steels are sensitive to localized corrosion. Stress corrosion cracking may occur if the materials are under mechanical stresses and the water contains impurities. If leaks in the tube wall were to occur, the exothermal reaction between alkali metal and water would then proceed in the liquid metal tubes and heat up the reaction mixture. The high pressure on the water side would provide the reaction with the reactant.

The hazards associated with such an eventuality in alkali metal technology have initiated intensive research work. Studies of reactions between liquid sodium and water or steam at a temperature of $\sim$400°C were started in early stages of the development of fast breeder reactors in several countries. Much less work has been performed on reactions of the other alkali metals with water and its vapor. The lithium–water reaction has been studied in small-scale tests, which have also been applied to reactions of potassium. The more technological studies of the sodium–water

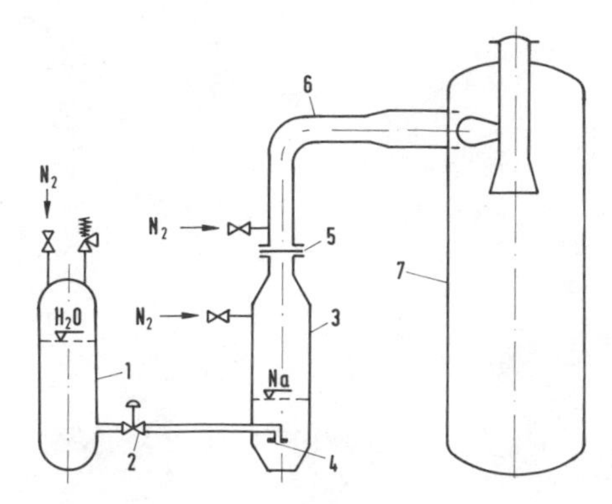

Figure 12.1. Test setup for sodium–water reactions in a vessel (schematic drawing after ref. 2).

reactions have used larger mock-up models of steam generators or superheaters,[1,2] an example of which is illustrated in Fig. 12.1.

12.1. The Reactions of Sodium with Steam

Liquid sodium reacts with steam of high pressure according to equation (1):

$$2Na_{(l)} + H_2O_{(g)} \rightarrow NaOH_{(s)} + NaH_{(s)} \tag{1}$$

This reaction takes place in an excess of alkali metal. The liquid metal does not form any layer of hydroxide or hydride which may separate the reactants and slow down the reaction. The equation indicates that the gaseous reactant is consumed and two solid products are formed. Thus, the reaction does not generate high pressure. The high energy liberated by the reaction, however, heats up the mixture and enhances the reaction rate. A second effect is the heating of the material constituting the tube wall. The heat evolution from the reaction is able to cause a local melting of the tube wall. Such a development of the reaction might cause severe damage to the reactor structure.

The reaction given by equation (1) develops higher pressure if the temperature reaches levels at which sodium hydride begins to decompose. The decomposition of hydride evolves hydrogen gas, and this gas evolution is the reason for a pressure buildup. The evolution of hydrogen, however, does not compensate the pressure decrease due to the consumption of steam. The evolution of hydrogen is an endothermal process. There is no danger of an explosive hydrogen reaction if the steam generator is still an intact barrier against contact with the atmosphere. The hydrogen chain reaction occurs, however, if air comes into contact with the reaction mixture. This chain reaction is much more violent than the sodium–steam reaction.

12.2. The Reaction of Sodium with Water

The reaction of hot liquid sodium with pressurized water is more dangerous than the reaction with steam. This reaction evolves gas out of a mixture of liquid reactants, as can easily be seen from equation (2).

$$Na_{(l)} + H_2O_{(l)} \rightarrow NaOH_{(s)} + NaH_{(s)} \tag{2}$$

$$NaH_{(s)} \rightarrow Na_{(l)} + \tfrac{1}{2}H_{2(g)} \tag{2a}$$

The heats of reaction of the individual steps occurring in the sodium–water reaction have been given by Bray.[1]

$$Na_{(l)} + H_2O_{(l)} \rightarrow NaOH_{(s)} + \tfrac{1}{2}H_2, \quad \Delta H^0_{298} = -145\ \text{kJ/mol} \tag{3a}$$

$$Na_{(l)} + H_2O_{(g)} \rightarrow NaOH_{(s)} + \tfrac{1}{2}H_2, \quad \Delta H^0_{298} = -185.4\ \text{kJ/mol} \tag{3b}$$

$$2Na_{(l)} + H_2O_{(g)} \rightarrow NaOH_{(s)} + NaH_{(s)}, \quad \Delta H^0_{298} = -195.3\ \text{kJ/mol} \tag{3c}$$

$$Na_{(l)} + \tfrac{1}{2}H_2 \rightarrow NaH_{(s)}, \quad \Delta H^0_{298} = -56.4\ \text{kJ/mol} \tag{3d}$$

$$Na_{(l)} + NaOH_{(s)} \rightarrow Na_2O + \tfrac{1}{2}H_2, \quad \Delta H^0_{298} = +10.7\ \text{kJ/mol} \tag{3e}$$

$$2Na_{(l)} + H_2O_{(g)} \rightarrow Na_2O_{(s)} + H_2, \quad \Delta H^0_{298} = -128.1\ \text{kJ/mol} \tag{3f}$$

$$2Na_{(l)} + NaOH_{(s)} \rightarrow Na_2O_{(s)} + NaH_{(s)}, \quad \Delta H^0_{298} = -45.7\ \text{kJ/mol} \tag{3g}$$

The development of hydrogen pressure caused by the sodium–water reaction has been measured in 1:6 scale tests in steam generator models, with provisions for the injection of water to simulate a leak in the tubes.[1] The effects of injection of 121 kg of water into the sodium in three seconds were investigated. The temperature of both media was maintained at 300°C. The pressure in the sodium was 0.5 bar and in the water 160 bars. It was shown that a maximum pressure of 29 bars developed at 50 ms after the initiation of the water injection. The sodium temperature locally rose to 930°C about two seconds after the injection had ceased. The temperature of the vessel and the bulk of the sodium reached 550°C. The pipework did not show any deformation due to these excursions of the pressure and the temperature. The reaction products—sodium hydroxide, sodium monoxide, and sodium hydride—remained inside the sodium pipes. Rock-like solid mixtures are the final products of the reaction. They are very difficult to remove by mechanical methods.

The sodium–water combustion has been studied by direct observation in a water container. The pressure changes can be detected by means of microphones. A fast film camera records the buildup and collapse of the flash. The camera has to be able to take 2000 frames in one second.[3] Weighed sodium samples have been introduced into the water surrounded by a wire mesh cage. The start of the reaction is vigorous. In its first phase, it produces smoke-filled bubbles with flashes of light. An explosion is observed after a delay of four seconds, if the amount of sodium is sufficient. The explosion is accompanied by an intense flash, the extension of which exceeds the region of the cage. The development of gaseous reaction products causes short separations of the reactants. The main reaction and explosion cannot start before the reactants come into contact again.

The temperature records indicate a temperature rise to a maximum value near 600°C. Before the end of the reaction, both temperature and pressure exhibit peak values. The maximum value of the temperature peak has been found to be close to 900°C.

The kinetics of sodium–water reactions are influenced by the geometry of the steam generator tubes.[2] Thus, modeling of such events requires a precise knowledge of the situation of a leak in tubes of specific sizes. The gas developed by the reaction is able to buffer the further process, as it separates the reacting liquids. The pressure in the sodium pipes, however, has sometimes exceeded the pressure of the water system. Special precautions to expand the sodium volume by means of burst disks into a pressure relief system have been found to be necessary. Modeling of the reaction in steam generator tubes has been performed and proved by Schnitker.[4]

A sudden rise of both temperature and pressure of a secondary sodium system indicates a leakage of steam generator tubes. Sensitive and fast measurements of these parameters can help to detect leaks before they cause harmful events. The analytical detection of hydrogen in sodium, however, is a much faster and more sensitive indicator of sodium–water reactions. Even the rise of hydrogen concentration in sodium can be disturbed, since reaction products have been able to close small leaks.

Several studies have been done in order to evaluate reaction rates of the sodium–water reaction. At the university of Nottingham a technique has been developed in which the metal surfaces are continuously renewed.[5] The formation of a surface layer separating the liquid metal from the vapor is suppressed by means of this technique. A glass cylinder, through which a stream of water passes downwards, is part of a small liquid metal circuit. The liquid metal is pumped through the circuit at a flow rate of one liter per minute. A sodium–potassium alloy (1:1) is used instead of molten sodium in order to keep the temperature low. The vigor of the reaction is limited by means of low gas pressure. Gas sampling through a quadrupole mass spectrometer has been employed to determine the changes in pressure and composition of the gas. An exposure time of less than one second has been found sufficient to allow the conclusion that the reaction is first order.

It has been demonstrated that the rate constant for the reaction of liquid NaK alloy with low-pressure water vapor depends on the temperature. At a temperature of 100°C its value is K (cm/s · Pa) $= 2.4 \times 10^{-3}$. The activation energy of this reaction is $E = 39$ kJ/mol. Water reacts orders of magnitude faster than hydrogen, and the reaction with water has the lowest activation energy.

The University of Nottingham group has also made studies on the behavior of a jet of the sodium–potassium alloy injected into water in a vessel made of glass.[5] Only a small part of the alloy injected into the water reacted at the orifice, and even the following layers of water did not consume the alkali metal at a temperature of 20°C. The two alkali metals were shown to react with the same reaction rate constant, though the reaction of potassium was faster than that of sodium. Figure 12.2 illustrates the

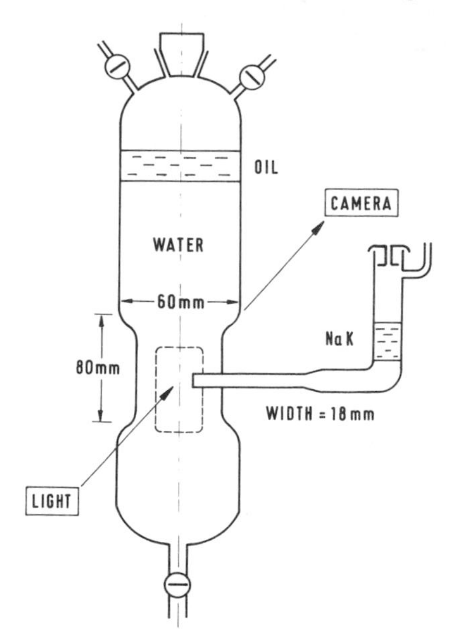

Figure 12.2. Apparatus utilized in the Nottingham laboratories to study the reactions of a jet of NaK alloy injected into water (schematic drawing after ref. 5).

principle of the apparatus used in the Nottingham laboratories. It has been found that sodium hydroxide formed in the sodium–water reaction is not stable and decomposes, as has been published by Whittingham *et al.*,[6] according to equation (4):

$$2Na_{(l)} + NaOH_{(l\ or\ s)} \xrightarrow{\text{fast}} Na_2O_{(s)} + NaH_{(s)} \quad (4)$$

$$Na_2O_{(s)} \xrightarrow{\text{slow}} (Na_2O)_{sol.} \quad (4a)$$

$$NaH_{(s)} \xrightarrow{\text{slow}} (NaH)_{sol.} \rightleftarrows \tfrac{1}{2}H_{2(g)} \quad (4b)$$

A contribution results from the reaction

$$Na_{(l)} + NaOH_{(l\ or\ s)} \xrightarrow{\text{fast}} Na_2O_{(s)} + \tfrac{1}{2}H_{2(g)} \quad (4c)$$

The results of several hydrogen gas pressure measurements in the atmosphere above the Na–NaOH mixtures indicate that only small portions of hydrogen are immediately evolved. Even at a temperature of 500°C the contribution of reaction (4c) remains small.

A steam generator leak in the Fermi reactor in December 1962 directed the interest of engineers to the behavior of sodium and water in the case of a small leak in the steam generator tubes. The event caused a drastic local thinning of the tube wall made of 2.25 Cr 1 Mo steel. The effect has been related to thermal effects of the reaction or to the sodium–water flames.

An experimental program[7,8] has shown that temperatures above 750°C were generated in the vicinity of the thinned section of the tube wall. It was obvious that a longer burning of that flame at the tube surface would have caused the formation of a hole due to the consumption of the material. Examinations of the material have shown that some spots around this hemispherically thinned area have been hardened. This hardening might be due to a precipitation of nonmetallic phases such as oxides in the neighborhood of the corroded area, while other regions have been annealed due to the high temperature of the flame.

The primary and secondary reaction products which are described by equations (2) and (4) cause secondary corrosion effects, the caustic corrosion. This process is relatively slow compared to the fast corrosion which is caused by the sodium–water flames.

The application of the sodium–water reaction for analytical purposes shows that the vigorous character of the reaction is suppressed by means of effective cooling and prevention of a hydrogen–oxygen chain reaction. The chain reaction does not occur if the reaction mixture is covered with an inert gas such as argon or nitrogen. The cooling has to keep the temperature of the water between 0 and 20°C in order to maintain the reaction at a low rate. It is necessary to add the alkali metal in small portions to restrain a local superheating.

12.3. The Reactions of Other Alkali Metals with Water

The vigor of the reactions occurring between alkali metals and water or steam intensifies with increasing atomic number of the metals. Several properties may favor the reactivity of the heavier alkali metals. The free energy of the reaction, however, has its highest value for the reaction between lithium and water according to

$$Li_{(s)} + H_2O_{(l)} \rightarrow LiOH_{(diss.)} + \tfrac{1}{2}H_{2(g)} \tag{5}$$

The free enthalpy of this reaction is $\Delta H = -508$ kJ/mol, and the values of the reactions of all the other alkali metals are somewhat lower.

The melting temperature of the metal seems to have an influence on the reaction rate, since molten metals cannot be protected by solid layers of reaction products. Lithium has the highest melting point and its reaction with water has the slowest rate. As pointed out in Chapter 2, the melting points of the other metals are much lower, with the melting points decreasing with increasing atomic number of the metal. Observations of the dissolution of alkali metals in cold water demonstrate that lithium and sodium remain in their solid state if the amount of water is large enough to remove the reaction heat. The other alkali metals melt on the surface of cold water and the reaction becomes more and more violent.

The removal of protecting layers takes place through the dissolution of the reaction products. The solubility of the hydroxides in the molten metals is very low. The solubility of the metal hydroxides in water, however, is high enough to play a role. Water dissolves only moderate amounts of lithium hydroxide. The hydroxides of the other alkali metals are more soluble, the effect intensifying with increasing atomic number.[9] The rate at which the hydroxides dissolve in water depends on their solubilities. Thus, hydroxide layers on potassium droplets in water dissolve much faster than layers of lithium hydroxide on solid pieces of this metal swimming on a water surface.

The reactions with water become more vigorous with alkali metals of higher atomic weight. The processes, however, can be maintained at rates which are not dangerous provided there is no contact with air, which may cause the hydrogen chain reaction. If the alkali metals are included in narrow slits, pores, or cracks of solid materials, the reaction might be stopped by hydrogen bubbles which separate the metal particles from the water. Hydroxide or oxide layers have the same protective character in such situations. It has been observed that such layers reduce the reaction rate to become very low or zero. A very slow growth of the hydroxide layers occurs and finally leads to the consumption of the alkali metals in such slits.

References

1. J. A. Bray, in: *Liquid Alkali Metals*, British Nuclear Energy Society, London, 1974, pp. 107–113.
2. K. Dumm, H. Mausbeck, and W. Schnitker, *Atomkernenergie 14*, 309–318 (1969).
3. R. N. Newman, A. R. Pugh, and C. A. Smith, in: *Liquid Alkali Metals*, British Nuclear Energy Society, London, 1974, pp. 85–91.
4. W. Schnitker, *Atomkernenergie 24*, 233–242 (1974).
5. A. B. Ashworth, M. G. Dowling, C. C. Addison, and R. J. Pulham, in: *Liquid Metal Engineering and Technology*, British Nuclear Energy Society, London, 1985, Vol. 3, pp. 1–6.
6. A. C. Whittingham, C. A. Smith, P. A. Simm, and R. J. Smith, in: *2nd Intern. Conf. on Liquid Metal Technology in Energy Production* (J. M. Dahlke, Ed.), National Techn. Information Service, Springfield, Va., 1980 (CONF-800401-P2), Vol. 2, 16/38.
7. R. A. Davies, J. A. Bray, and J. M. Jones, in: *Alkali Metal Coolants*, International Atomic Energy Agency, Vienna, 1967, pp. 263–272.
8. H. O. Muenchow and W. L. Chase, in: *Alkali Metal Coolants*, International Atomic Energy Agency, Vienna, 1967, pp. 273–289.
9. R. C. Weast (Ed.), *Handbook of Chemistry and Physics*, 60th Ed., The Chemical Rubber Co., Cleveland, Ohio, 1979.

13

Tribology in Alkali Metals

The tribology of solid materials wetted by liquid alkali metals is related to the friction generated on account of the relative motion of one of the surfaces against the other one. Thus, tribology deals with the friction, wear, and self-welding of materials in contact with each other in a liquid metal environment. Conventional lubricants are not compatible with alkali metals. The wetting liquids act, therefore, as lubricating liquids. Their lubricating properties are, however, poor compared to lubricants normally used in technology. The materials used for the movable parts of some essential components, for instance, valves, must have an adequate tribologic behavior for the construction of bearings. These materials have to guarantee the closure of the valves by means of complete contacting. Self-welding, on the other hand, has to be excluded.

The fuel elements of a sodium-cooled fast breeder reactor form hexagonal lattice. Since the single elements with their wrapper tubes have to be kept movable, the friction between their contacting surfaces should remain low even after a long period of exposure to high-temperature sodium so that the elements can be pulled out using a limited amount of force. Hence, the development of sodium-cooled fast breeder reactors has led to the initiation of research and development programs to study the tribology of several materials in the liquid metal environment over a large range of temperatures. Parameters which have been studied are the purity of the alkali metal, the composition and physical properties of the materials, and the physical conditions in the sliding system, such as the force with which the surfaces are pressed on each other, the velocity of the relative motion, and the original surface roughness. Such programs have been undertaken in the seventies in the United States,[1] the United Kingdom,[2] Japan,[3] and the Federal Republic of Germany.[4]

The tribologic properties of materials are dependent on their composition and physical properties. Thus, the austenitic stainless steels usually used as structural and fuel element cladding materials in fast breeder reactors are relatively ductile, and they tend to be self-welded under conditions of sufficiently high temperature, pressure, and sodium purity.[5] On the other hand, several hard-faced alloys based on nickel or cobalt do not show this tendency. The coating of material surfaces using composite materials containing nonmetallic phases such as oxides or carbides protects them against self-welding and reduces wear problems as well. Even the *in situ* formation of nonmetallic surface layers has been shown to minimize self-welding and wear.[6]

13.1. Wear under Liquid Sodium

The evaluation of materials of high wear resistance needs an experimental setup in which a pair of specimens is kept in relative motion under sodium and pressed together by means of a well-defined force. The motion is a torsion of a couple of pins pressed on a disk of the partner material. The results of such tests are influenced by the materials of the couples. Hence, it is important which material is used for the pins and which for the disks.[4] The phenomena caused by wear are weight losses due to rubbing or the deformation of the surfaces with negligible weight losses.

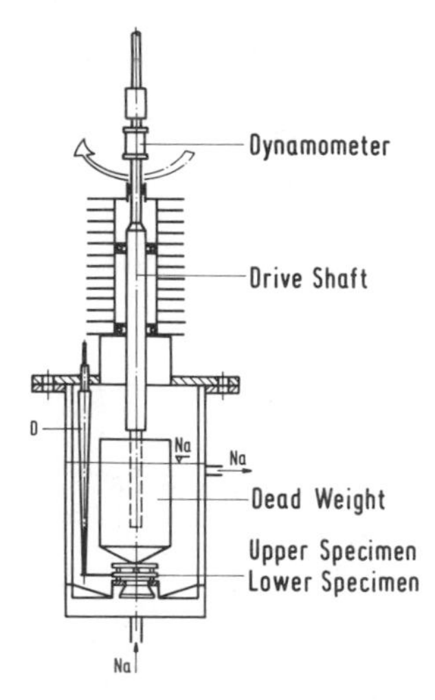

Figure 13.1. Test section for rotating relative movement (after ref. 7; reproduced with the permission of Kernforschungszentrum Karlsruhe GmbH).

Since the temperature and the oxygen potential of the environment of flowing sodium have to be well defined in these tests, the wear test sections have to be connected to a sodium loop. A loop provided with a cold trap and an oxygen meter keeps the liquid metal environment at the surfaces of the specimens within the desired range of parameters.

Figure 13.1[7] shows a pin-on-disk configuration of a test section designed for rotating or intermittent relative motion. The motion of the drive shaft connected to the pin holder (upper specimen) is measured through a dynamometer. The pressure of the pins on the disk (lower specimen) is exerted by means of a dead weight. Sodium enters the vessel at the bottom and leaves it through the side wall fairly well above the specimens. The lower specimen (disk) is prevented from simultaneous rotation by means of a bending lever (D in Fig. 13.1). This also serves to measure the friction force at the disk, while the torque of the drive shaft is measured through a torsion cell. It has been shown that the results are significantly influenced by the stiffness of the test section. The wear caused at the pins and the disk is measured by

- change in length of the cylindrical pins
- surface roughening of the annular disk
- change in weight of both sliding partners

Results of such tests using hybrid pairings of disk (Inconel 718, tungsten carbide in nickel matrix) and pins (Colmonoy 6 and 56, Tantalum, FeCr 50, Stellite 1 and 6H) are shown in Fig. 13.2.[7] The wear rates measured for several couples of specimens made of identical materials are shown in Fig. 13.3. The figure demonstrates that disks and pins are affected to different degrees of severity. There are some material couples which are sufficiently resistant against wear under sodium.

DISK MATERIAL	Wear Penetration ← → Change of Length 1000 100 10 μm 10 100 1000	PIN MATERIAL
Inconel 718		Colm. 6
		Tantal
		Colm. 56
		FeCr 50
		Stell. 6H
WC (Ni)		Stell. 1
		Colm. 6

Figure 13.2. Wear under sodium of hybrid pairings of materials (after ref. 7; reproduced with the permission of Kernforschungszentrum Karlsruhe GmbH).

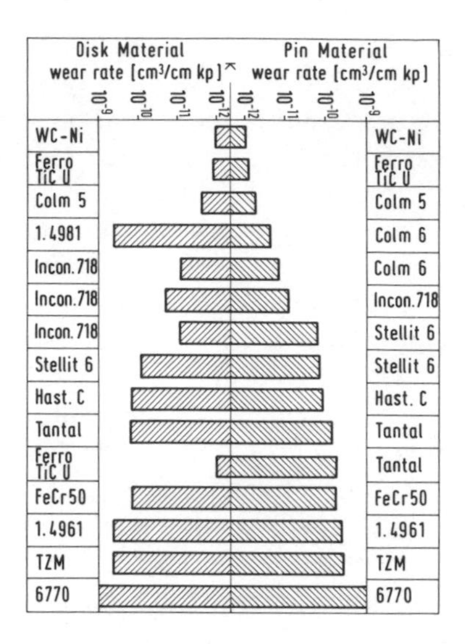

Figure 13.3. Wear rates under sodium of different couples of materials (after ref. 7; reproduced with the permission of Kernforschungszentrum Karlsruhe GmbH).

Some of the wear-resistant materials which are of different composition than the structural materials are applied for surface protection, for instance, in the form of pads welded on the surfaces of components which they have to protect.

The comparison of the wear of Stellite and Colmonoy under flowing sodium at 280 to 540°C has shown that the wear depth which is developed on the surface of Colmonoy is lower than on Stellite. The values measured are independent of the sliding velocity in the tests under sodium. However, the sliding mode (oscillation or continuous sliding in one direction) has an influence. Oscillating motion leads to higher values of the wear depth and increased roughening of the surfaces. The wear rate under sodium is higher at 540 than at 450°C.[3]

Chromium carbide (5-μm Cr_3C_2 starting powder in a binder of metallic character) is used as a wear-resistant coating. Coating layers of 0.075 to 0.125 mm thickness are deposited on stainless steel surfaces by means of detonation-gun coating or plasma spray techniques and finally treated to get a surface finish. Such materials are compatible with liquid sodium under conditions of a fast breeder reactor core and do not suffer irradiation damage to a degree which interferes with their protective character. Wear testing in flowing sodium revealed their high wear resistance even under the influence of the liquid metal environment.[8]

Wear is also caused by rolling of a cylinder on flat bearings when the two are moved against each other under the liquid metal.[9] For axially loaded, angular contact bearings, rotation is incompatible with pure rolling on both races, and wear arises from the resulting ball spin. The torque and instantaneous wear rate of such a bearing are similar functions of the integral over the contact areas of the product of contact pressure and radius from the ball spin axis. A better estimate of wear coefficient could be obtained by relating the average torque, the average wear, the initial torque, and the initial wear where the conditions are known. Tests have been made at 400°C in sodium. The results indicate specific wear rates of the order of 10^{-15} $m^3/N \cdot m$. These values are comparable with data from conventional sliding tests.

13.2. Friction in High-Temperature Sodium

The importance of low-friction behavior of contacting materials has been emphasized by reactor designers. The movement of adjacent surfaces requires low friction coefficients. For in-core applications, materials such as load pads on fuel as well as control rod ducts must have low friction coefficients (below 0.4 during refueling and below 0.9 during operation). These low values have to be maintained even after long dwell times and in spite of corrosion and mass transfer or irradiation damage. Low-friction couples have to be selected from pairs of materials showing low wear depth.

The friction coefficient μ is related to the friction forces tangential to the contact area F_T and the load normal to the motion F_N[7]

$$\mu = \frac{F_T}{F_N} \tag{1}$$

The friction coefficient depends on the nature of the motion. Measurements at the onset of a movement yield the breakaway friction coefficient, μ_b, whereas measurements during continuous sliding bring out the dynamic friction coefficient, μ_d. The maximum value in a continuous or intermittent movement is called the static friction coefficient, μ_s. Breakaway friction appears after dwell periods and is, therefore, of importance for the procedure of fuel element unloading.

The experimental setups used for the measurement of friction coefficients differ from those used in wear tests. The principle of such a test section, which is also connected to a sodium loop, is shown in Fig. 13.4. Three plates are pressed together by mechanical forces. The central one is moved by means of vibrational forces developed through a hydraulic testing machine. The shaft is connected to the flange of the test section through a

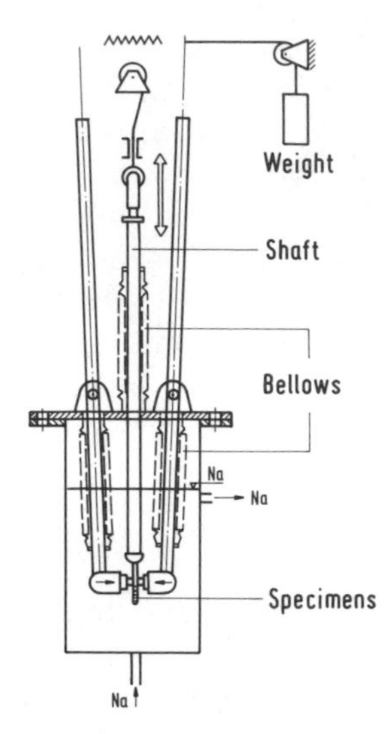

Figure 13.4. Test section for translatory oscillating movement to evaluate friction coefficients (after ref. 7; reproduced with the permission of Kernforschungszentrum Karlsruhe GmbH).

bellow. The two outer plates are pressed on the central one by means of a system of two levers and loads. An example of the records of friction coefficients obtained in such a test setup is shown in Fig. 13.5. The different coefficients are also defined in this record. It is obvious from results of measurements in different laboratories that the experimental equipment has a strong influence on the measured values of the friction coefficients. Wild *et al.*[7] have related the influence of the testing equipment to the stiffness of the loading system.

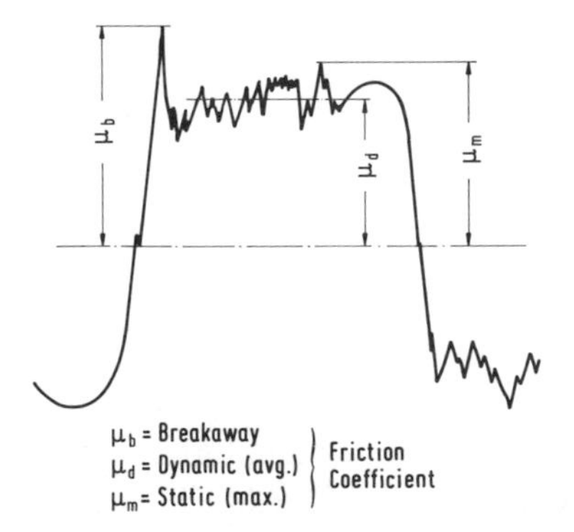

Figure 13.5. Typical record of friction coefficients as measured in oscillating friction tests (after ref. 7; reproduced with the permission of Kernforschungszentrum Karlsruhe GmbH).

The friction studies of the wear-resistant materials Inconel 718 and Stellite do not yield the necessary low values of the friction coefficients.[1,7] A nickel-based alloy which seems to remain in the low-friction region at a temperature of 538°C is Tribaloy 700 (manufactured by E. I. du Pont de Nemours & Co. with a composition of 50 wt% Ni, 32 wt% Mo, 15 wt% Cr, and 3 wt% Si).[10] The carbon content of this alloy does not play an important role. Tribaloy 700 has to be applied as weld-deposited plating produced by means of detonation-gun or plasma spray techniques. The alloy shows a low friction coefficient $\mu < 0.4$ up to a sodium temperature of 600°C at a contact pressure of 2070 kPa. Tests at 625°C show that the friction coefficient slightly exceeds this limit and increases with an increasing number of rubbing cycles. Dwell periods of one week do not show a significant influence on the sliding behavior of this material.

Inconel 718, however, tends to give higher values of friction coefficients even at lower temperatures of the liquid metal environment. The scattering of the results of such tests is explained by the fact that the material loses grains from the bulk alloy which are deformed between the rubbing surfaces.[10] Stellite 6 hard facings on several substrate materials exhibit the same tendency. The dynamic friction coefficients tend to increase above the $\mu = 0.4$ limit at temperatures below 400°C and above 600°C. Since microstructural changes occur when the temperature is greater than 480°C for an exposure time of more than 1000 hours, these effects may be the reason for the occurrence of high friction coefficients at high temperature.

The grain boundary effects of sodium exposure on Inconel 718 are definitely suppressed by an aluminide coating. This surface treatment generates a diffusion coating of nickel aluminide, NiAl, which drastically reduces the friction coefficients compared to unprotected Inconel 718.[1] The pack-cementation process is used to produce such a diffusion coating.

The coating of materials with chromium carbide in metallic binder also reduces the friction coefficients in the molten alkali metal medium. The best results have been obtained with a coating of Cr_3C_2 powder in Tribaloy 700 binder plated on the substrate by means of the D-gun process.[8] Though all materials show increasing values of the friction coefficients as the sodium temperature increases, the nonmetallic coatings seem to promise sufficiently low values up to a temperature of 600°C. The coatings discussed here are compatible with sodium even at this high temperature level. The effects of high-temperature sodium above this level might be due to a degradation of the coatings caused by sodium corrosion reactions. High-temperature sodium ($T > 650$°C) extracts aluminum from nickel aluminide coatings, thus reducing the protective character of the coating layer. The carbide coatings, however, seem to suffer a loss of sur-

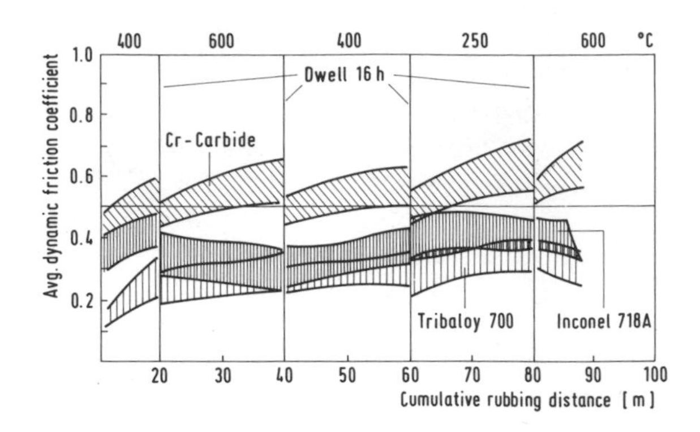

Figure 13.6. Average dynamic friction coefficients of coating materials in tests under sodium (after ref. 7; reproduced with the permission of Kernforschungszentrum Karlsruhe GmbH).

face carbide grains due to the corrosive action of sodium at this temperature level.

Wild *et al.*[7] have also tested the friction behavior of these coatings in liquid sodium. Their results, which are shown in Fig. 13.6, indicate that Tribaloy 700 is superior to the two nonmetallic coatings. The chromium carbide coating exhibits friction coefficients above the limit of $\mu = 0.4$, while aluminized Inconel 718 and Tribaloy 700 remain within the limit in the temperature range 400 to 600°C.

13.3. Self-Welding in High-Temperature Sodium

Self-welding of materials in contact with each other in the presence of liquid sodium at a sufficiently high temperature is a serious tribologic problem. For instance, self-welding of stainless steel specimens occurs even in the absence of forces pressing them together in high-temperature sodium corrosion tests. Flanges which are exposed to sodium above a temperature of 650°C and with a certain pressure are often self-welded. This effect limits their use since they cannot be opened after the tests. The same phenomenon might be observed with valves, if they are used in the high-temperature region of a sodium loop.

The tests to evaluate the self-welding behavior of pairs of materials include a dwell time of the pairs of specimens in flowing or stagnant sodium and the post-exposure measurement of the breakaway force which is necessary to separate the welded specimens from each other. The testing devices are similar to the wear test equipment. They allow the pressing of a column or a pair of specimens on each other under the liquid metal. In order to get a high specific pressure, a small contact area is pressed on a flat, polished counter area. The test section is provided with purified and heated sodium from a sodium loop. Important parameters of self-welding tests are the temperature of the sodium environment, the dwell time, the contact pressure, the size of the contact area, the purity of sodium, the surface quality, and the thermomechanical pretreatment of the materials. It has been shown that the results of the breakaway measurements depend on the contact area, the contact pressure, the time of contact under pressure, the surface roughness, and the temperature of the test section.(5) Metallographic studies reveal that a real diffusion bonding is formed during the dwell time under sodium. It seems that the influence of sodium is due to its reducing action on the contacting surfaces and its capacity to dissolve and precipitate components of alloys, thus transporting them from one specimen to the other. Figure 13.7 exhibits the self-welding zone of a pair of two stainless steel (X8 CrNiMoNb 16 16) specimens which have been contacted more than 300 hours at 700°C under sodium with a contact pressure of 550 kgf. The metallographic cross section clearly shows that the separating line between the two specimens has disappeared due to the self-welding process. The boundary zone has undergone a recrystallization. Self-welding zones in pairs of different steels have an intermediate diffusion zone.(11)

Several materials including ferritic steels with 2.25 and 9 wt% chromium content, unstabilized and stabilized austenitic steels, nickel- and cobalt-based alloys, and refractory alloys (such as the molybdenum-based alloy TZM) have been tested in two laboratories,(5,11) while the Westinghouse Advanced Reactors Division has only tested the self-welding behavior of Inconel 718 and Stellite.(10)

The ferritic steels show the strongest tendency to be self-welded under sodium. The rate constant for self-welding of a pair of 2.25 Cr–1 Mo steel surfaces is higher by a factor of six than that for an austenitic steel pair (such as, for instance, type AISI 316).(11) Hard materials such as the alloys Inconel 718 and Stellite do not show a tendency to be self-welded under liquid metals to such a degree. The alloy TZM is not self-welded even after pairs of specimens of this material have been pressed under sodium at 700°C.(5) The content of titanium and zirconium in this molybdenum-based alloy seems to be the reason for its resistance against self-welding. These components form stable oxides in sodium, even if it is of excellent purity, and this prevents the diffusion of metal atoms from onc partner of the couple to the other. It has been shown that stainless steel can also be protected against self-welding by the deposition of zirconium vapor on its surface.(6)

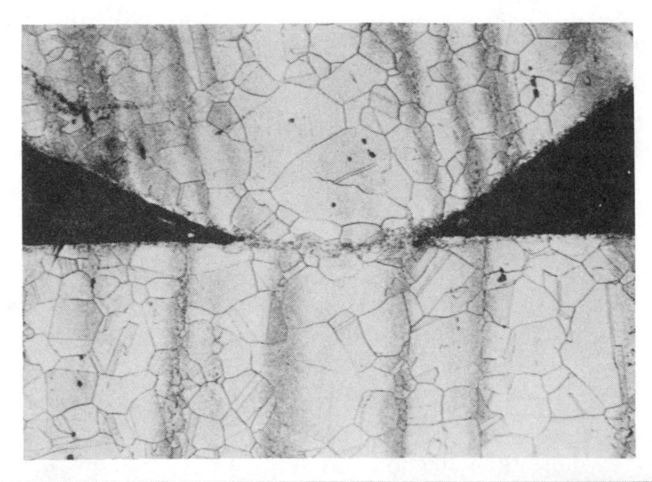

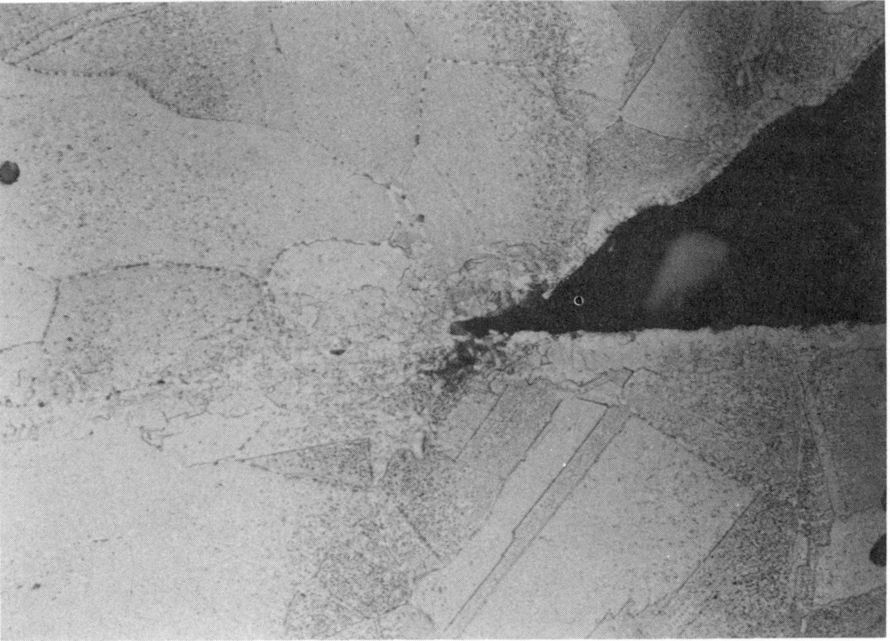

Figure 13.7. Metallograph of the self-welded area of a pair of stainless steel specimens contacted 336 h at 700°C and a pressure of 550 kgf under sodium (after ref. 5).

The activation energy for the self-welding of pairs of ferritic 2.25 Cr–1 Mo steel and of the austenitic steel AISI 316 has been measured to be $\Delta H_{act} = 22$ kcal/mol (=90.6 kJ/mol) for both pairs.

13.4. The Influence of Sodium Chemistry on Tribology

Since the formation of surface oxides is able to suppress self-welding of several material pairs, it is obvious that the chemical activity of oxygen in sodium and the activity of oxygen-gettering metallic components in the alloys control the tendency for self-welding. If the alloys contain considerable amounts of getter elements such as zirconium, the oxygen dissolved in the liquid metal plays only a minor role. The affinity of zirconium is high enough to consume nearly all the oxygen present in the alkali metal. However, if chromium is the oxide-forming component of an alloy, the oxygen potential in the liquid metal is of importance for the formation of a protecting surface oxide layer. Thus, a field of a low-friction regime might be defined in the temperature versus oxygen concentration system of coordinates. The extension of this regime depends on the chemical activity of chromium in the surface of the solid material, as is shown in Fig. 13.8.

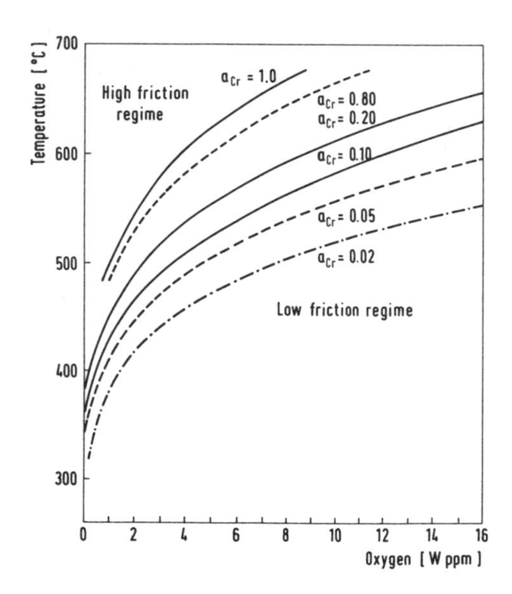

Figure 13.8. High- and low-friction regimes of steels contacted under liquid sodium as influenced by the temperature and oxygen concentration (and $NaCrO_2$ formation) (after ref. 1; reproduced with the permission of Mr. Hoffman).

Table 13.1. Composition of Wear-Resistant Materials

Alloy component	Inconel 718	TZM	Stellite 6 H	Colmonoy[a] 6	Tribaloy 700
			Content in wt%		
C	0.05	0.025	1.0	2.6	low
Cr	19	—	27	16.5	15
Ni	53	—	—	70	50
Fe	19	—	—	—	—
Al	0.7	—	—	—	—
Ti	1	0.5	—	—	—
Nb	5	—	—	—	—
Mo	3	99.4	—	—	32
Zr	—	0.1	—	—	—
W	—	—	4.5	—	—
Co	—	—	67.5	—	—
B	—	—	—	3.8	—

[a] Contains Fe + Si + C ~10 wt% max.

It is easy to explain the different self-welding tendencies of ferritic and austenitic steels in liquid sodium on the basis of this diagram. One has also to conclude that the leaching of chromium out of austenitic steels as a result of sodium corrosion increases the susceptibility to self-welding. The fact that the friction coefficients of most of the materials tested exceed the limit of $\mu = 0.4$ at a temperature above 600°C might also be related to the diagram in Fig. 13.8. The oxygen concentration in the sodium circuits equipped with cold traps which are used in technology is mostly kept at a level below 10 wppm. The steels or alloys with a chemical activity of chromium $a_{Cr} = 0.2$ leave the low-friction regime just above 600°C at this level of oxygen concentration. The figure indicates that surface plating with chromium should also be beneficial for the tribology of material surfaces in liquid alkali metals.

Protective thin oxide layers are also involved in the use of nickel aluminide coatings or zirconium deposits. The protective character of aluminide is due to the chemical activity of aluminum in this layer. Leaching of this element decreases its chemical activity, thus causing a decrease of the protecting effect. However, the high oxygen affinity of aluminum is the reason why oxide is formed as long as aluminum is present in the surface zone. Zirconium has an extremely low solubility in liquid sodium. Its oxide formed in the presence of liquid sodium is a black compound of very fine grains which covers the metal surface and is not easily removed by flowing sodium.[12]

Besides the chemical activities of dissolved nonmetals which interfere with the tribology of materials, the dissolved metals are also able to show interference effects. Thus the chemical activity of iron in liquid sodium causes the precipitation of crystals on the surfaces of Stellite. These crystals contain iron as their main component. Such deposits cover the entire surfaces of low-friction materials such as Stellite and Inconel 718. The high iron content of the sliding surfaces is the reason that the behavior moves into the high-friction regime as is indicated in Fig. 13.8. Additionally, the softness of the deposition layers causes high friction coefficients as well. Rubbing of such Stellite surfaces completely deforms the iron crystals which play the role of a very ineffective lubricant.(13)

Since carbides are the alternative class of protective coatings on sliding material surfaces, the carbon chemistry of sodium should be of similar importance. The carbide Cr_3C_2 is not very stable against sodium of low carbon potential. However, the precipitated phases do not directly equilibrate with the carbon activity of the liquid metal. Kinetic factors are more important than in the case of oxides. The dissolution rates of carbon or carbon compounds in sodium are very low. Thus, the loss of carbide due to mechanical rubbing might be faster than the chemical dissolution.

13.5. Tribology in Other Alkali Metals

The tribology of materials in alkali metals other than sodium has not yet been studied. The increasing interest in the technological use of lithium and potassium makes a knowledge of the influence of these molten metals on the tribology of structural materials necessary. The state of work on friction and self-welding in sodium of different chemical compositions allows some conclusions to be drawn with respect to lithium and potassium.

Liquid lithium reduces chromium oxides and does not form complex lithium chromium oxides. Stainless steels and nickel-based alloys are, therefore, more or less in the high-friction regime when contacted in liquid lithium environment even at relatively low temperature. The low-friction regime requires the presence of metallic elements which form very stable oxides. Nickel aluminide coatings seem to be compatible with lithium, and they might be able to improve the situation of surfaces in sliding contact in molten lithium. The deposition of zirconium may also cause the formation of stable surface oxide layers. However, zirconium seems to be more soluble in lithium than in sodium.

The chemical stability of carbides against lithium does not differ from the behavior in sodium. Thus, carbide coatings in a binder metal which does not contain high amounts of nickel might be a suitable alternative solution for the friction problems. Fusion reactor technology has still to initiate studies on tribology in liquid lithium as well as in the eutectic lithium–lead alloy.

The tribology in molten potassium might be less problematic than in sodium. However, results of studies have not been published so far. The free energy of formation of potassium oxide is less negative than that of sodium oxide. Thus, surface oxides on solid materials, such as passivation layers on stainless steels, are more stable against potassium than against sodium, and the low-friction regime might be ruled by the existence of such surface layers. Complex oxides do not seem to play the same role in tribology in liquid potassium as they do in liquid sodium. This tendency might also be found in systems with the heavy alkali metals rubidium and cesium.

References

1. R. N. Johnson, R. C. Aungst, N. J. Hoffman, M. G. Cowgill, G. A. Whitlow, and W. L. Wilson, in: *Intern. Conf. on Liquid Metal Technology in Energy Production* (M. H. Cooper, Ed.), National Techn. Information Service, Springfield, Va., 1976 (CONF-760503-P1), pp. 122–130.
2. H. W. Roberts, Friction and Wear Behaviour of Sliding Bearing Materials in Sodium Environments at Temperatures up to 600°C, TRG Report 1269, United Kingdom Atomic Energy Authority, 1966.
3. S. Kanoh, S. Mizobuchi, and H. Atsumo, in: *Intern. Conf. on Liquid Metal Technology in Energy Production* (M. H. Cooper, Ed.), National Techn. Information Service, Springfield, Va., 1976 (CONF-760503-P1), pp. 153–159.
4. E. Wild and K. J. Mack, in: *Intern. Conf. on Liquid Metal Technology in Energy Production* (M. H. Cooper, Ed.), National Techn. Information Service, Springfield, Va., 1976 (CONF-760503-P1), pp. 131–137.
5. F. Huber and K. Mattes, in: *Intern. Conf. on Liquid Metal Technology in Energy Production* (M. H. Cooper, Ed.), National Techn. Information Service, Springfield, Va., 1976 (CONF-760503-P1), pp. 177–183.
6. F. Huber, K. Mattes, and H. Weinhold, Verfahren zur Verhinderung des Selbstverschweissens von metallischen Werkstoffen in flüssigem Natrium, German patent no. 2 342 850, 1975.
7. E. Wild, K. J. Mack, and M. Gegenheimer, Liquid Metal Tribology in Fast Breeder Reactors, Report KfK 3738, Kernforschungszentrum Karlsruhe, 1984.
8. G. A. Whitlow, L. Wilson, T. A. Galioto, R. L. Miller, S. L. Schrock, N. J. Hoffman, J. J. Droher, and R. N. Johnson, in: *Intern. Conf. on Liquid Metal Technology in Energy Production* (M. H. Cooper, Ed.), National Techn. Information Service, Springfield, Va., 1976 (CONF-760503-P1), pp. 138–144.
9. C. S. Campbell, in: *Intern. Conf. on Liquid Metal Technology in Energy Production* (M. H. Cooper, Ed.), National Techn. Information Service, Springfield, Va., 1976 (CONF-760503-P1), pp. 160–166.

10. N. J. Hoffman, J. J. Droher, J. Y. Chang, T. A. Galioto, R. L. Miller, S. L. Schrock, G. A. Whitlow, W. L. Wilson, and R. N. Johnson, in: *Intern. Conf. on Liquid Metal Technology in Energy Production* (M. H. Cooper, Ed.), National Techn. Information Service, Springfield, Va., 1976 (CONF-760503-P1), pp. 167–176.
11. N. Yokota and S. Shimoyashiki, in: *Liquid Metal Engineering and Technology*, British Nuclear Energy Society, London, 1984, Vol. 1, pp. 99–103.
12. H. U. Borgstedt, G. Frees, and G. Drechsler, *Werkst. Korros. 21*, 568–573 (1970).
13. H. Schneider, in: *Second Intern. Conf. on Liquid Metal Technology in Energy Production* (J. M. Dahlke, Ed.), National Techn. Information Service, Springfield, Va., 1980 (CONF-800401-P1), 3–52.

14

Alkali Metal Batteries

Batteries based on alkali metals are capable of giving high specific power and energy as well as efficient and long-term performance. The theoretical specific energy (W_m) for a given cell reaction is given by

$$W_m = \frac{-\Delta G}{\Sigma m} \tag{1}$$

where ΔG is the Gibbs free energy change for the reaction and Σm is the sum of the weights of the reactants. The thermodynamic reversible potential E is related to ΔG by the equation

$$E = \frac{-\Delta G}{nF} \tag{2}$$

where n is the number of electrons involved in the cell reaction and F is the Faraday constant. As alkali metals are the most electropositive of elements, cell reactions in which they combine with halogens or chalcogens would result in high theoretical EMF. Batteries based on the lighter alkali metals would also have high specific energy. Thus a lithium–fluorine cell would have a cell voltage of 6.05 V and a specific energy of 6250 Wh/kg. No such cell has so far been developed. Lithium–chlorine and lithium–sulfur cells would have $E = 3.98$ and 2.18 V and $W_m = 2500$ and 2550 Wh/kg, respectively. The theoretical specific energies given above do not take into account the weights of electrolyte, cell housing, and terminal. When this is done the specific energy values are lowered by a factor of 3 to 6. Even then the performance of alkali metal batteries would be far superior to that of conventional batteries.[1,2]

Alkali metals make ideal electrodes. They are good conductors of electricity. In the liquid form they wet the surfaces of the electrolyte, thus minimizing internal resistances. The liquid electrode is particularly useful at high current densities, which necessitate the rapid diffusion of the active electrochemical species. Liquid metal electrodes operate without polarization under such conditions.

There is considerable interest today in high-performance batteries having high energy and power densities. Primary batteries with long life and high specific energy are required by the electronics industry for use in electronic watches, calculators, pacemakers, etc. Military applications demand batteries with long shelf life. In particular, pyrotechnically activated "thermal batteries" are of special interest to defense installations, for example, as power sources in guided missiles. Lithium-based primary batteries have emerged as the major contenders for all these applications.

The main interest in secondary batteries having high specific energy and power is centered around their applications in energy storage and vehicle traction. Energy storage batteries are of interest to the electric utility industry, which has to cope with fluctuating demand for energy. As Birk[(3)] has shown, an electric utility company in the United States makes use of three types of power-generating units. Almost half of the capacity is derived from base-load plants, which make use of coal, gas, or nuclear fuel. They operate continuously and meet about 70% of the total energy demand throughout the year. Intermediate plants operating during the daily peak hours from morning till evening provide about 30% of the capacity. The peaking power, representing less than 10% of the system capacity, comes from units which produce only 2 to 3% of the annual energy consumption. Thus the energy from these units is very costly. However, if energy storage batteries are available, they can be charged during slack hours and the stored power drawn during the peak hours. Figure 14.1, taken from ref. 3, illustrates this possibility.

For electric traction the battery must not only have high specific energy, but also high specific power. For adequate acceleration of a small vehicle, an instantaneous power of 50 kW is required, while a 300-km range demands a capacity of about 50 kWh. In both applications the battery must have long cycle life and high efficiency in order to be economical. A ten-year life demanded in energy storage corresponds to about 2500 cycles of charge–discharge. A five-year life is considered adequate for vehicle applications. Table 14.1, taken from ref. 4, lists the performance goals.

The most promising high-specific-energy cells are based on alkali metals. Lithium-based primary cells are already in use in electronic consumables. Sodium–sulfur and lithium–sulfur batteries are the leading can-

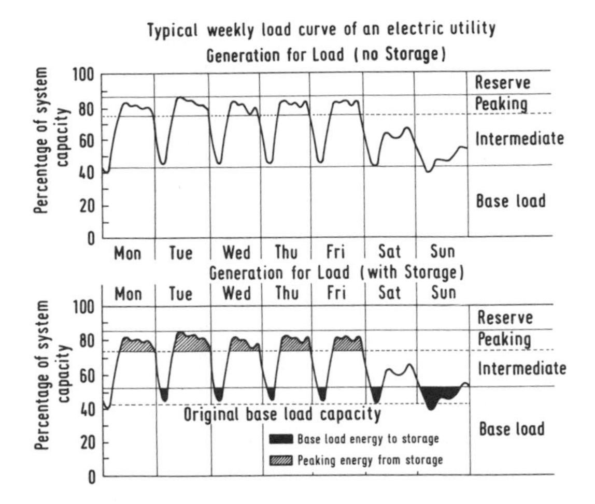

Figure 14.1. Typical weekly load curve of an electric utility (after ref. 3; reproduced with the permission of Plenum Press).

didates for energy storage and vehicle traction. Figure 14.2, taken from ref. 5, compares the specific power–specific energy curves for alkali metal batteries with those for conventional batteries (calculated) for electric traction applications.

In this chapter we will first discuss the primary batteries and then the secondary batteries.

Table 14.1. Performance Goals for Electric Vehicle and Off-Peak Energy Storage Battery Cells

	Electric vehicle propulsion	Off-peak energy storage
Normal discharge cycle (h)	2–5	5–10
Normal charge cycle (h)	5–10	5–10
Specific energy ($Wh \cdot kg^{-1}$)	100–165	100–165
Specific power ($W \cdot kg^{-1}$)	135[a]	
Cell current density ($A \cdot cm^{-2}$)		
Peak	0.4–0.5	
Normal	0.05	0.05
Watt-hour efficiency (%)	70	80
Cycle life	1000	1500

[a] Value at 50% discharge of the cell capacity.

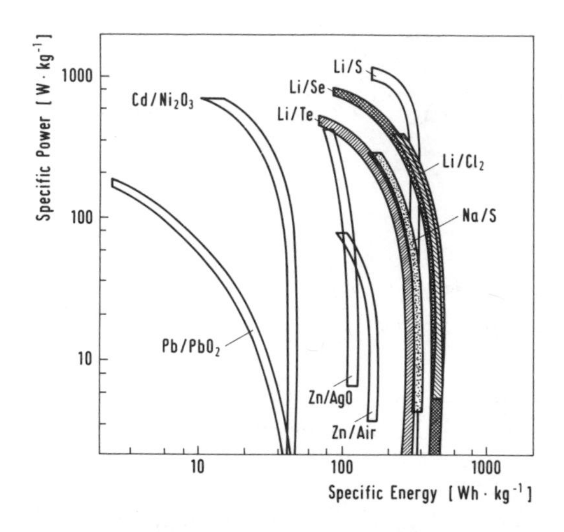

Figure 14.2. Specific power–specific energy curves for batteries (after ref. 5; reproduced with the permission of the American Association for the Advancement of Science).

14.1. Lithium-Based Primary Batteries

Primary batteries which are small in size but have a high energy output are constantly in demand in the electronics industry. Silver oxide and mercury oxide cells have been popularly used for this purpose. However, batteries based on lithium are rapidly replacing these cells because of their higher voltage and higher specific energy. Lithium systems which have been under development include Li–SO_2, Li–$SOCl_2$, Li–Cl_2, Li–Ag_2CrO_4, Li–$(CF)_n$, and Li–MnO_2. All of them give a cell voltage of about 3 V. These batteries generally use a lithium anode, a liquid electrolyte based on an organic or inorganic solvent, and a cathode active material such as thionyl chloride. Batteries based on solid electrolytes have also been developed, especially for special applications such as in pacemakers.

Perhaps the most popular among these primary cells is the lithium–thionyl chloride cell.[6,7] The system consists of a lithium anode and a carbon cathodic collector with a glass fiber separator between them. Thionyl chloride acts as cathode active material and a solvent for a suitable electrolyte salt, e.g., $LiAlCl_4$. The cell reaction is given by the equation[8]

$$4Li + 2SOCl_2 \rightarrow S + SO_2 + 4LiCl \quad (3)$$

These cells are capable of giving high energy densities in the range of 700 to 800 Wh/dm^3. The schematic of a cell shown in Fig. 14.3 is taken from ref. 7.

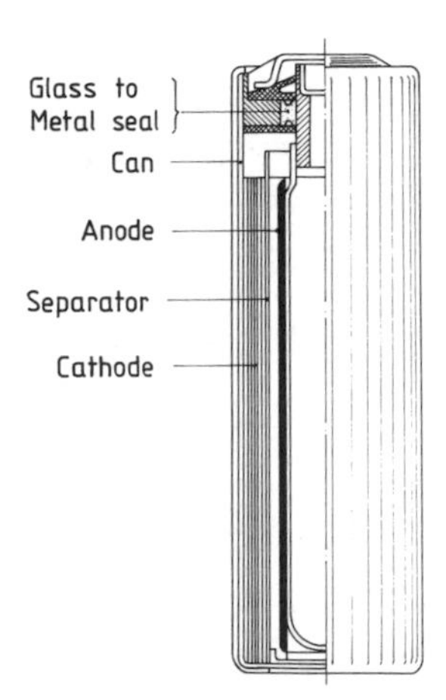

Figure 14.3. Schematic of a lithium–thionyl chloride cell (after ref. 7; reproduced with the permission of Academic Press, Inc., Orlando, Florida, USA).

These cells have a good shelf life. One of the disadvantages of the Li–$SOCl_2$ cells has been a voltage delay when put to use after extended periods of storage. This delay arises as a result of the formation of a protective layer of LiCl on the lithium anode which is in contact with thionyl chloride. In fact, this protective layer accounts for the stability of the lithium anode in such electrolytes. Improved electrolytes (such as $AlCl_3$ in $SOCl_2$ neutralized by Li_2O) seem to alleviate this problem.

The lithium–chlorine cell[5,9] makes use of a liquid lithium anode contained in a porous stainless steel wick, a porous carbon electrode, and a molten lithium chloride electrolyte separating them. Chlorine gas is fed to the cathode from external storage. The reactions at the anode and cathode are, respectively,

$$Li \rightarrow Li^+ + e^- \tag{4}$$

$$\tfrac{1}{2}Cl_2 + e^- \rightarrow Cl^- \tag{5}$$

The electrolyte melting point being 609°C, the operating temperature of the cell is high (650°C). While this battery is capable of high specific power and high specific energy, it appears to have problems relating to impurity accumulation in the cathode and the collection of the electrolysis products on recharge. Corrosion is also a problem.

The lithium–sulfur dioxide cells make use of a lithium anode and sulfur dioxide dissolved in a suitable medium (e.g., acetonitrile–LiBr) which serves as both cathode active material and electrolyte.[10] A suitable cathode collector completes the system. A Li–$(CF)_n$ system in which polycarbonmonofluoride [$(CF)_n$] serves as the cathode and an appropriate organic-based solution (e.g., $LiBF_4$ dissolved in propylene carbonate) as the electrolyte has been described by Fukuda and Iijima.[11] The cell reaction

$$nx\text{Li} + (\text{CF}_x)_n \rightarrow nx\text{LiF} + n\text{C} \qquad (6)$$

gives a cell potential of about 3 V, and an energy density of 1400 to 2000 Wh/kg has been claimed to be theoretically possible. A lithium–silver chromate battery composed of a lithium anode, a silver chromate cathode, and a liquid electrolyte formed of a lithium perchlorate solution in propylene carbonate has been developed for medical uses (e.g., pacemakers).[12] It is said to be capable of practical energy densities of 600 W/dm^3 and a long storage life. Recently Ikeda[13] has described a lithium–manganese dioxide cell which is claimed to have the advantages of high energy density and low cost.

Solid-electrolyte primary cells based on lithium have been developed for low-power, long-life applications such as implantable cardiac pacemakers. A typical cell consists of a lithium anode, a lithium iodide electrolyte, and a suitable iodine or sulfur compound as cathode.[14,15] Schneider *et al.*[14] have described a cell in which a charge transfer complex of iodine and poly(2-vinylpyridine) is used as the cathode. Lithium iodide electrolyte is formed *in situ*. These cells are reported to be very reliable and to have a long life of more than six years. Solid electrolytes having high lithium ion conductivities are not known and hence solid-state batteries suitable for high-power-density applications have not been developed.

Thermal batteries are primary cell systems based on molten salt electrolytes that are inert until brought into use by the ignition of a charge of pyrotechnic. On being activated, the battery supplies power for periods ranging from a few seconds to an hour depending upon the design and the rate at which it is discharged. Lithium-based thermal batteries have become very attractive in recent years because of their much superior performance compared to the earlier calcium-based cells.

The most common electrolyte used in thermal batteries is a eutectic mixture of lithium and potassium chlorides with a melting point of 352°C.[16,17] Lithium or its alloy with aluminum or silicon constitutes the anode. An FeS_2 cathode completes the cell. The internal construction of a typical thermal battery shown in Fig. 14.4 is taken from ref. 16. The anode in a metal cup is separated from a pellet containing FeS_2 and the electrolyte by a pure electrolyte plus binder pellet. This ensures that lithium does not react directly with FeS_2. Such three-pellet units are separated by

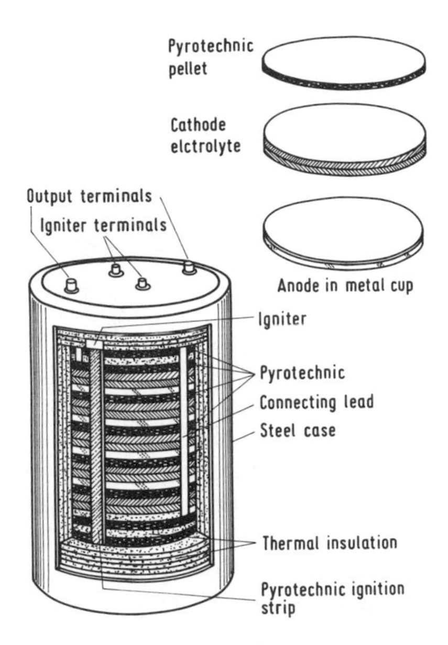

Figure 14.4. Thermal battery (after ref. 16; reproduced with the permission of Academic Press, Inc., Orlando, Florida, USA).

pyrotechnic pellets (usually Fe–$KClO_4$). Other thermal batteries studied include Li–Cl_2[18] and Li|LiCl–KCl|$CaCrO_4$|Fe.[19,20]

Thermal batteries have an almost indefinite shelf life. Upon activation, the electrolyte and the anode become molten and the battery starts giving an output in a fraction of a second. Lithium-based thermal batteries (typically giving about 30 V and 0.5 A for one hour)[21] outperform all other types of thermal batteries.

14.2. The Lithium–Sulfur Battery

The lithium–sulfur cell was first introduced by Cairns and Shimotake in 1969.[22] Their early work with molten lithium and sulfur as electrodes and a eutectic melt of lithium chloride–potassium chloride as electrolyte met with only limited success. This led to the development of a cell with solid electrodes. Lithium alloys[23] are now used as anode and iron sulfides[24] as cathode, leading to the cell:

$$\text{Li–Al}_{(s)} \text{ or Li–Si}_{(s)}|\text{LiCl–KCl}_{(l)}|\text{FeS}_{2(s)} \text{ or FeS}_{(s)}$$

When the cell discharges, several intermediate electrochemical reactions take place with the eventual formation of lithium sulfide and α-iron. The overall cell reaction is

$$2Li + FeS \rightarrow Li_2S + Fe \tag{7}$$

or

$$2Li + FeS_2 \rightarrow Li_2S + FeS \tag{8}$$

or

$$4Li + FeS_2 \rightarrow 2Li_2S + Fe \tag{9}$$

depending on the extent to which the reaction is allowed to proceed. Reaction (8) gives an open-circuit voltage (OCV) of 2.06 V and reaction (7), 1.66 V. The operating temperature of the cell is between 380 and 475°C. When lithium–aluminum alloy is used as the anode with FeS as cathode, the OCV is only 1.33 V and the theoretical specific energy of the system comes down to 460 Wh/kg from 2600 Wh/kg for the lithium–sulfur couple.

A solid alloy of lithium is preferred over liquid lithium as the anode because the latter attacks the ceramic insulator as well as the electrolyte. The reaction with the electrolyte would release volatile potassium:

$$Li_{(l)} + KCl_{(l)} \rightarrow LiCl_{(l)} + K_{(g)} \tag{10}$$

Lithium–aluminum alloy is easy to prepare. Over a considerable composition range up to 47 at% (~18 wt%) of lithium, the system is a two-phase mixture of a solid solution of lithium in aluminum and the compound β-LiAl. Therefore, compositional variations do not affect the open-cell voltage. The anode operates reversibly in the composition range of 4 to 18 wt% lithium. Lithium–silicon alloy has the attractive feature that it would reduce the weight of the anode. It is possible to incorporate up to 95 at% of lithium in a lithium–silicon alloy.

Sulfur itself cannot be used as the cathode as it dissolves in the molten electrolyte. This sulfur-containing electrolyte then reacts with the anode, leading to self-discharge of the cell. Therefore, iron sulfides were chosen as cathode materials. While FeS_2 gives a higher cell voltage it is also more corrosive, thus necessitating the use of more expensive collector materials like molybdenum. It also tends to lose sulfur on cycling. FeS is not only less corrosive but cheaper as well. However, it leads to a lower cell voltage. Selenium and tellurium can be used in place of sulfur as cathode. While the use of a heavier chalcogen lowers the specific energy of the cells, its higher electronic conductivity reduces the amount of current collector in the cathode.[25]

A prismatic cell which has evolved out of the developmental work at

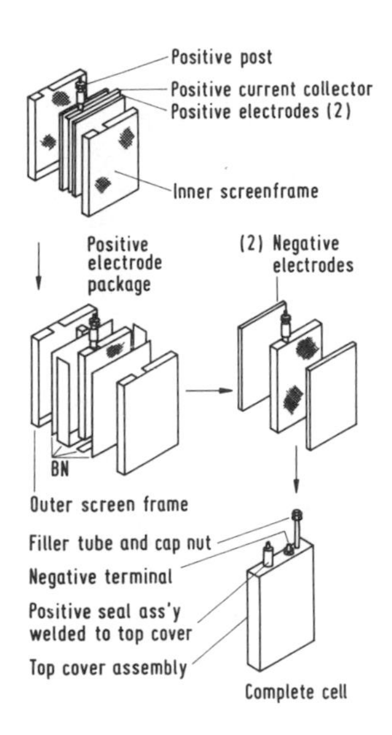

Figure 14.5. Lithium–sulfur battery (after ref. 27; reproduced with the permission of Academic Press, Inc., Orlando, Florida, USA).

Argonne National Laboratory is described by Walsh and Shimotake.[26] A three-plate cell being developed by Askew *et al.*[27] is shown in Fig. 14.5. It consists of a positive electrode consisting of FeS (or $FeS + Fe + Li_2S$, corresponding to a partially charged composition) mixed with the electrolyte and pressed into two half-electrode plates with a current collector between them. This electrode is strengthened by wire mesh on either side and separated on either end from the lithium–aluminum negative electrodes by boron nitride separators. Askew *et al.* reported a specific energy of 90 to 100 Wh/kg. More recently Barlow *et al.*[28] have reported the development of a low-cost cell design incorporating a powder separator and an immobilized ternary halide electrolyte. Efforts are being made to improve the cycle life of the cells so that they become useful for energy storage applications.

14.3. All-Solid-State Lithium Battery

All-solid-state secondary batteries based on lithium are a newer development.[29] Here the electrolyte is a composite of poly(ethylene oxide) (PEO) and a lithium salt such as lithium trifluoromethane sulfonate

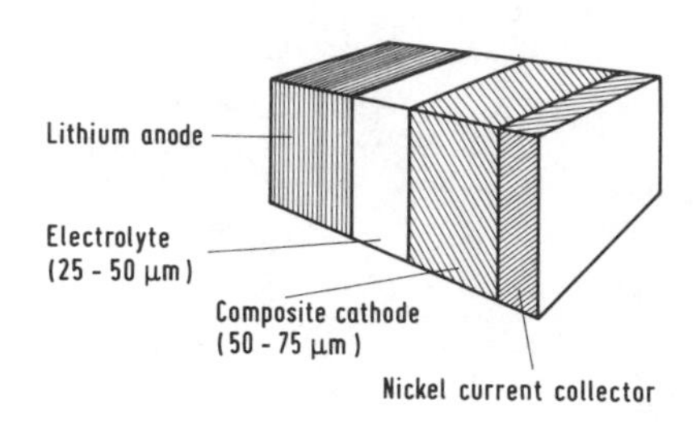

Figure 14.6. Principle of an all solid-state lithium cell.

($LiCF_3SO_3$). The active materials are a lithium anode and an insertion cathode, V_6O_{13}. The cathode is also made as a porous composite material from V_6O_{13}, PEO, and carbon. Figure 14.6 illustrates the construction of the cell. The cell reaction is

$$Li_x^+ + xe^- + V_6O_{13} \underset{}{\overset{\text{discharge}}{\rightleftarrows}} Li_xV_6O_{13} \qquad (0 < x < 8;\ V^{+4}/V^{+5} \rightarrow V^{+3}) \tag{11}$$

The cell operates in the temperature range 120–140°C and gives a cell voltage of 2.4 volts. The lithium ion conductivity of the electrolyte is optimized by adjusting the salt content in it. Even though the ionic conductivity of the polymer-based electrolyte is lower than that of some inorganic electrolytes, the voltage drop across the electrolyte can be minimized by the use of very thin electrolyte sheets. This is because pore-free polymer membranes of 25–50 μm thickness can be readily fabricated. Because the total thickness of the type of cell shown in Fig. 14.6 is very small (0.1 mm), very compact batteries of high capacity and voltage can be fabricated out of such cells. Various possible cell configurations can be designed, which makes the battery quite flexible. A practical specific energy of greater than 200 Wh/kg and a life of more than 150 cycles appear possible. Ruggedness, ease of fabrication, and safety are the other advantages claimed for this battery.

14.4. The Sodium–Sulfur Battery

The sodium–sulfur battery makes use of molten sodium and sulfur as anode and cathode, respectively. These electrodes are separated by a sodium-ion-conducting ceramic material called beta alumina. The cell operates at a temperature of 300 to 350°C.

During the cell discharge cycle, sodium gets oxidized at the anode by the half-cell reaction

$$Na \rightarrow Na^{+} + e^{-} \tag{12}$$

Sodium ions migrate through the beta alumina and reduce sulfur at the cathode according to the half-cell reaction

$$e^{-} + Na^{+} + xS \rightarrow NaS_x \tag{13}$$

Sodium forms a number of polysulfides. In the first phase of the discharge of the cell, Na_2S_5 is formed, giving an open-circuit voltage of 2.08 V. This voltage remains constant so long as the two liquid phases, Na_2S_5 and S, coexist. As the reaction proceeds, all sulfur gets consumed and then lower polysulfides get formed (Na_2S_x with $5 > x > 3$, see Fig. 8.5). Now the cell voltage decreases with x. When x falls below 3, solid Na_2S_2 starts precipitating. Since this is undesirable from the point of view of reversibility and beta alumina life, cell discharge is stopped at a cell potential of 1.78 V. Now the cell can be recharged by reversing the above reactions by applying the appropriate potential.

The key component in the sodium–sulfur cell is the solid electrolyte, beta alumina. In fact, it was the discovery of sodium ion conduction in this material by Yao and Kummer in 1967[30] that led to the development of this cell. This electrolyte can be prepared in two forms:

(1) β-alumina with a composition of $Na_2O \cdot 11Al_2O_3$ (Fig. 14.7)
(2) β''-alumina with a composition of $Na_2O \cdot 5.33Al_2O_3$.

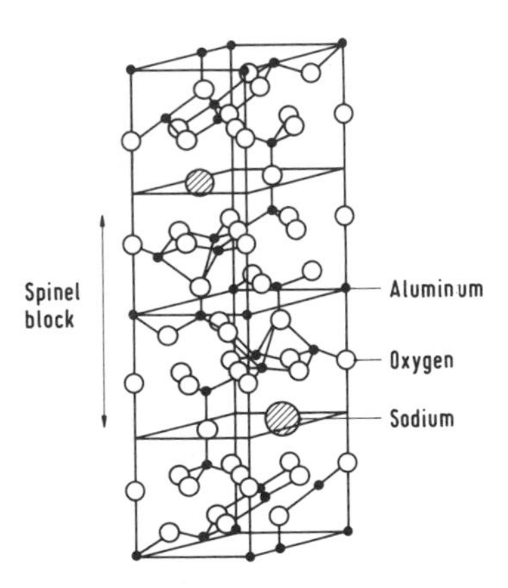

Figure 14.7. Crystal structure of the beta alumina electrolyte.

β-Alumina is built up of spinel-structure blocks with the formula $Al_{11}O_{16}^{+}$ which are separated by layers containing sodium and oxygen. These layers act as two-dimensional conducting planes for sodium ions. In both compounds the conduction planes are separated by 11.3 Å. The β-alumina unit cell is made up of two spinel blocks (2×11.3 Å) while β''-alumina has three blocks (3×11.3 Å) in its unit cell. The spinel block in β''-alumina has a cation vacancy which is compensated by three extra sodium ions in the conduction plane. Thus, β''-alumina has the higher ionic conductivity and hence is the popular choice for making storage cells. However, it is metastable in the pure form and hence requires additives to stabilize it. About 1 to 2% Li_2O plus MgO appears optimum. β-Alumina electrolytes of the desired shapes are produced by techniques such as electrophoretic deposition or isostatic pressing followed by sintering at 1600 to 1700°C.[31] Fine grain structure of the ceramic is important to obtain good strength and superior ionic conductivity. Electronic conductivity must, of course, be absent.

Figure 14.8 gives the schematic of a typical cell, taken from ref. 31. In this case sodium is contained inside the beta alumina tube whereas sulfur with graphite fiber current collector randomly distributed in it forms the cathode on the outside of the electrolyte tube. Sulfur is contained in a stainless steel or mild steel container with appropriate surface treatment.

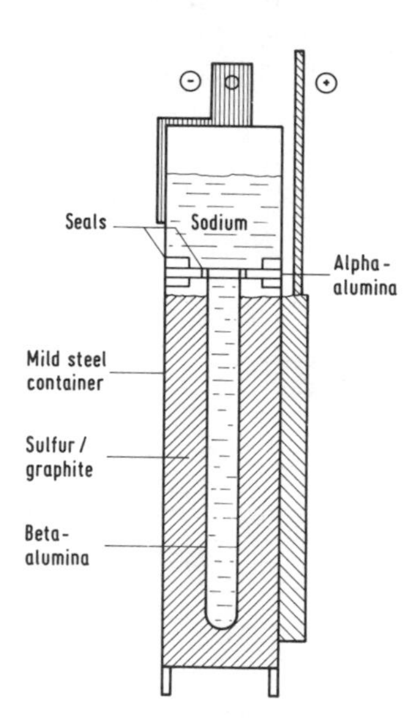

Figure 14.8. Schematic of the sodium–sulfur cell (after ref. 31; reproduced with the permission of the author).

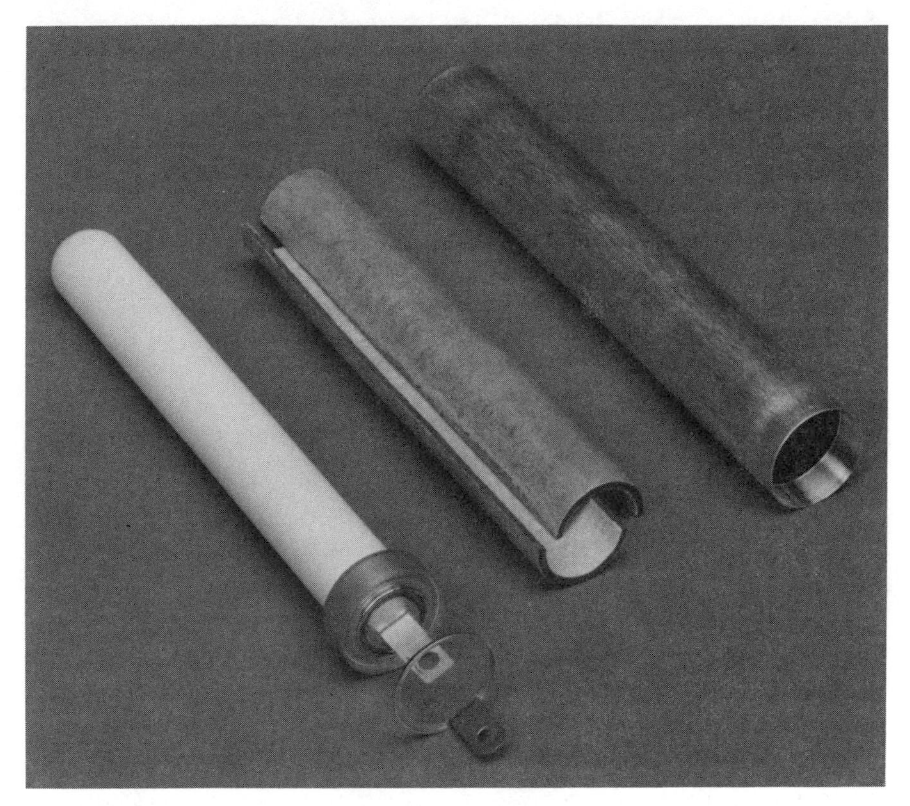

Figure 14.9. Sodium–sulfur cell of the battery designed for vehicle traction by Brown, Boveri & Cie., Heidelberg, Germany. (Courtesy Dr. Mennicke.)

The beta alumina tube is sealed with glass to an alpha alumina ring which serves as the electrical insulator between the two electrodes. The alpha alumina ring is in turn sealed to the outer container through ceramic-to-metal seals. Alternate designs reverse the electrodes—sulfur inside the electrolyte tube and sodium outside it.[32,33]

Ford Motor Company, General Electric Company, Brown, Boveri & Cie., Chloride Silent Power Ltd., British Rail, and Shangai Institute of Ceramics are among the companies developing sodium–sulfur batteries. Batteries with specific energies of 150 Wh/kg have been built and tested. Cycle life of 500 charge–discharge cycles seems to be typical. Figure 14.9 shows the components of a battery developed at Brown, Boveri & Cie., Heidelberg, for vehicle traction. Cars and pickup trucks based on such batteries are expected to be in trial production shortly.

Further developments are required to improve the reliability and life of the cell and its safety. Reliability and life depend largely on the quality of the beta alumina electrolyte. The impact of the reactions at the electrode surface on the charge–discharge processes at high current densities are also receiving attention. Corrosion of the container by sulfur and polysulfides also deserves further study. Liquid sodium does not pose any serious problem. However, the safety of the system must be demonstrated under conditions when container integrity is breached, in order to gain public acceptance, especially for vehicle traction. Cell containers are being designed to virtually eliminate such risks. A major demonstration of these batteries expected in the mid-eighties will be crucial to their acceptance.

References

1. J. P. Gabano (ed.), *Lithium Batteries*, Academic Press, New York, 1983.
2. E. Peled and H. Yamin, in: *Power Sources 8* (J. Thompson, Ed.), Academic Press, New York, 1981, pp. 101–115.
3. J. R. Birk, in: *Superionic Conductors* (G. D. Mahan and W. L. Roth, Eds.), Plenum Press, New York, 1976, pp. 1–14.
4. E. C. Gay, D. R. Vissers, N. P. Yao, F. J. Martino, T. D. Kaun, and Z. Tomczuk, in: *Power Sources 6* (D. H. Collins, Ed.), Academic Press, New York, 1977, pp. 735–749.
5. E. J. Cairns and H. Shimotake, *Science 164*, 1347–1355 (1969).
6. A. N. Dey and P. Bro, in: *Power Sources 6* (D. H. Collins, Ed.), Academic Press, New York, 1977, pp. 493–510.
7. J. P. Gabano and G. Gelin, in: *Power Sources 8* (J. Thompson, Ed.), Academic Press, New York, 1981, pp. 3–15.
8. J. C. Bailey and J. P. Kohut, in: *Power Sources 8* (J. Thompson, Ed.), Academic Press, New York, 1981, pp. 17–26.
9. D. A. J. Swinkels, *J. Electrochem. Soc. 113*, 6–10 (1966).
10. P. Bro, R. Holmes, N. Marincic, and H. Taylor, in: *Power Sources 5* (D. H. Collins, Ed.), Academic Press, New York, 1975, pp. 703–712.
11. M. Fukuda and T. Iijima, in: *Power Sources 5* (D. H. Collins, Ed.), Academic Press, New York, 1975, pp. 713–728.
12. G. Gerbier, G. Lehmann, P. Lenfant, and J. P. Rivault, in: *Power Sources 6* (D. H. Collins, Ed.), Academic Press, New York, 1977, pp. 483–492.
13. H. Ikeda, in: *Power Sources 8* (J. Thompson, Ed.), Academic Press, New York, 1981, pp. 63–76.
14. A. A. Schneider, W. Greatbatch, and R. Mead, in: *Power Sources 5* (D. H. Collins, Ed.), Academic Press, New York, 1975, pp. 651–659.
15. J. R. Rea, L. H. Barnette, C. C. Liang, and A. V. Joshi, Electrochem. Soc., Extended Abstracts, Vol. 79-2, Abstract no. 3 (1979).
16. A. Attewell and A. J. Clark, in: *Power Sources 8* (J. Thompson, Ed.), Academic Press, New York, 1981, pp. 285–303.
17. J. G. Searcy, R. K. Quinn, and H. J. Saxton, in: *Power Sources 9* (J. Thompson, Ed.), Academic Press, New York, 1983, pp. 563–578.
18. F. M. Bowers and J. H. Ambrus, in: *Power Sources 5* (D. H. Collins, Ed.), Academic Press, New York, 1975, pp. 595–602.

19. G. G. Bowser, D. Harney, and F. Tepper, in: *Power Sources 6* (D. H. Collins, Ed.), Academic Press, New York, 1977, pp. 537–547.
20. M. D. Baird, A. J. Clark, C. R. Feltham, and L. J. Pearce, in: *Power Sources 7* (J. Thompson, Ed.), Academic Press, New York, 1979, pp. 701–712.
21. R. K. Quinn, D. E. Zurawski, and N. R. Armstrong, in: *Power Sources 8* (J. Thompson, Ed.), Academic Press, New York, 1981, pp. 305–321.
22. E. J. Cairns and H. Shimotake, *J. Sulph. Inst. 5*, 5 (1969).
23. N. P. Yao, W. J. Walsh, H. Shimotake, J. D. Arntzen, and P. A. Nelson, Electrochem. Soc., Extended Abstract no. 12 (1973).
24. D. R. Vissers, Z. Tomczuk, and R. K. Steunenberg, *J. Electrochem. Soc. 5*, 121 (1974).
25. R. K. Steunenberg, in: *Intern. Conf. on Liquid Metal Technology in Energy Production* (M. H. Cooper, Ed.), National Techn. Information Service, Springfield, Va., 1976, (CONF-760503-P2), Vol. 2, pp. 485–491.
26. W. J. Walsh and H. Shimotake, in: *Power Sources 6* (D. H. Collins, Ed.), Academic Press, New York, 1977, pp. 725–733.
27. B. A. Askew, P. V. Dand, L. P. Eaton, T. W. Olszanski, and E. J. Chaney, in: *Power Sources 8* (J. Thompson, Ed.), Academic Press, New York, 1981, pp. 337–355.
28. G. Barlow, P. Dand, and B. Askew, in: *Power Sources 9* (J. Thompson, Ed.), Academic Press, New York, 1983, pp. 543–561.
29. B. C. Tofield, R. M. Dell, and J. Jensen, UKAEA Report AERE-r 11261, Harwell, 1984.
30. Y. F. Y. Yao and J. T. Kummer, *J. Inorg. Nucl. Chem. 29*, 2453–2479 (1967).
31. K. W. Browall, *Industrial Research & Development*, 108 (Dec. 1981).
32. A. D. Seeds, M. L. Weight, and M. D. Hames, in: *Power Sources 8* (J. Thompson, Ed.), Academic Press, New York, 1981, pp. 323–335.
33. I. Wynne Jones, in: *Liquid Metal Engineering and Technology*, British Nuclear Energy Society, London, 1985, Vol. 3, pp. 317–321.

Index